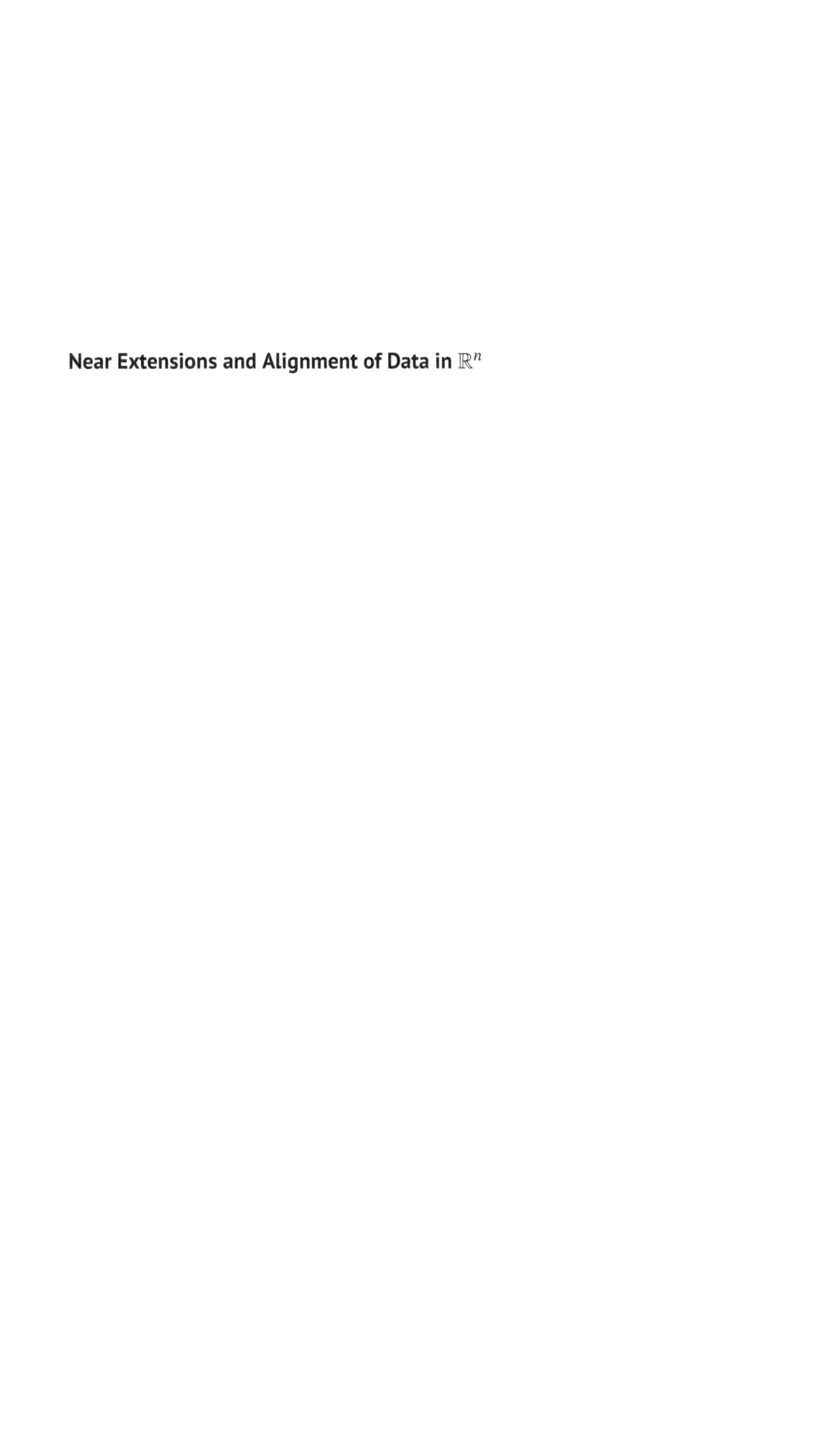

Near Extensions and Alignment of Data in $\mathbb{R}^n$

Near Extensions and Alignment of Data in $\mathbb{R}^n$

Whitney extensions of near isometries, shortest paths, equidistribution, clustering and non-rigid alignment of data in Euclidean space

Steven B. Damelin
Ann Arbor, MI, USA

This edition first published 2024

Registered Offices
John Wiley & Sons, Inc., 111 River Street, Hoboken, NJ 07030, USA
John Wiley & Sons Ltd, The Atrium, Southern Gate, Chichester, West Sussex, PO19 8SQ, UK

For details of our global editorial offices, customer services, and more information about Wiley products visit us at www.wiley.com.

Wiley also publishes its books in a variety of electronic formats and by print-on-demand. Some content that appears in standard print versions of this book may not be available in other formats.

Library of Congress Cataloging-in-Publication Data

Names: Damelin, Steven B., author.
Title: Near extensions and alignment of data in R^n : Whitney extensions of near isometries, shortest paths, equidistribution, clustering and non-rigid alignment of data in Euclidean space / Steven B. Damelin.
Description: Hoboken, NJ : John Wiley & Johns, 2024. | Includes bibliographical references and index.
Identifiers: LCCN 2023032013 | ISBN 9781394196777 (hardback) | ISBN 9781394196791 (adobe pdf) | ISBN 9781394196807 (epub) | ISBN 9781394196814 (ebook)
Subjects: LCSH: Mathematical analysis | Geometry, Analytic. | Rigidity (Geometry) | Nomography (Mathematics) | Euclidean algorithm. | Isometrics (Mathematics)
Classification: LCC QA300 .D325 2024 | DDC 516.3–dc23/eng/20231002
LC record available at https://lccn.loc.gov/2023032013

Cover Design: Wiley
Cover Image: Courtesy of the Author

Set in 9.5/12.5pt STIXTwoText by Integra Software Services Pvt. Ltd, Pondicherry, India
Printed and bound by CPI Group (UK) Ltd, Croydon, CR0 4YY

C9781394196777_031123

To my family

Contents

Preface *xiii*
Overview *xvii*
Structure *xix*

1 Variants 1–2 *1*
1.1 The Whitney Extension Problem *1*
1.2 Variants (1–2) *1*
1.3 Variant 2 *2*
1.4 Visual Object Recognition and an Equivalence Problem in $\mathbb{R}^d$ *3*
1.5 Procrustes: The Rigid Alignment Problem *4*
1.6 Non-rigid Alignment *6*

2 Building ε-distortions: Slow Twists, Slides *9*
2.1 c-distorted Diffeomorphisms *9*
2.2 Slow Twists *10*
2.3 Slides *11*
2.4 Slow Twists: Action *11*
2.5 Fast Twists *13*
2.6 Iterated Slow Twists *15*
2.7 Slides: Action *15*
2.8 Slides at Different Distances *18*
2.9 3D Motions *20*
2.10 3D Slides *21*
2.11 Slow Twists and Slides: Theorem 2.1 *23*
2.12 Theorem 2.2 *23*

3 Counterexample to Theorem 2.2 (part (1)) for card (E)> d *25*
3.1 Theorem 2.2 (part (1)), Counterexample: $k > d$ *25*
3.2 Removing the Barrier $k > d$ in Theorem 2.2 (part (1)) *27*

4 Manifold Learning, Near-isometric Embeddings, Compressed Sensing, Johnson–Lindenstrauss and Some Applications Related to the near Whitney extension problem *29*
4.1 Manifold and Deep Learning Via c-distorted Diffeomorphisms *29*
4.2 Near Isometric Embeddings, Compressive Sensing, Johnson–Lindenstrauss and Applications Related to c-distorted Diffeomorphisms *30*
4.3 Restricted Isometry *31*

5 Clusters and Partitions *33*
5.1 Clusters and Partitions *33*
5.2 Similarity Kernels and Group Invariance *34*
5.3 Continuum Limits of Shortest Paths Through Random Points and Shortest Path Clustering *35*
5.3.1 Continuum Limits of Shortest Paths Through Random Points: The Observation *35*
5.3.2 Continuum Limits of Shortest Paths Through Random Points: The Set Up *36*
5.4 Theorem 5.6 *37*
5.5 p-power Weighted Shortest Path Distance and Longest-leg Path Distance *37*
5.6 p-wspm, Well Separation Algorithm Fusion *38*
5.7 Hierarchical Clustering in $\mathbb{R}^d$ *39*

6 The Proof of Theorem 2.3 *41*
6.1 Proof of Theorem 2.3 (part(2)) *41*
6.2 A Special Case of the Proof of Theorem 2.3 (part (1)) *42*
6.3 The Remaining Proof of Theorem 2.3 (part (1)) *45*

7 Tensors, Hyperplanes, Near Reflections, Constants (η, τ, K) *51*
7.1 Hyperplane; We Meet the Positive Constant η *51*
7.2 "Well Separated"; We Meet the Positive Constant τ *52*
7.3 Upper Bound for Card (E); We Meet the Positive Constant K *52*
7.4 Theorem 7.11 *52*
7.5 Near Reflections *52*
7.6 Tensors, Wedge Product, and Tensor Product *53*

8 Algebraic Geometry: Approximation-varieties, Lojasiewicz, Quantification: (ε, δ)-Theorem 2.2 (part (2)) *55*
8.1 Min–max Optimization and Approximation-varieties *56*
8.2 Min–max Optimization and Convexity *57*

9 Building ε-distortions: Near Reflections *59*
9.1 Theorem 9.14 *59*
9.2 Proof of Theorem 9.14 *59*

10 ε-distorted diffeomorphisms, *O(d)* and Functions of Bounded Mean Oscillation (BMO) *61*
10.1 BMO *61*
10.2 The John–Nirenberg Inequality *62*
10.3 Main Results *62*
10.4 Proof of Theorem 10.17 *63*
10.5 Proof of Theorem 10.18 *66*
10.6 Proof of Theorem 10.19 *66*
10.7 An Overdetermined System *67*
10.8 Proof of Theorem 10.16 *70*

11 Results: A Revisit of Theorem 2.2 (part (1)) *71*
11.1 Theorem 11.21 *71*
11.2 η blocks *74*
11.3 Finiteness Principle *76*

12 Proofs: Gluing and Whitney Machinery *77*
12.1 Theorem 11.23 *77*
12.2 The Gluing Theorem *78*
12.3 Hierarchical Clusterings of Finite Subsets of $\mathbb{R}^d$ Revisited *81*
12.4 Proofs of Theorem 11.27 and Theorem 11.28 *82*
12.5 Proofs of Theorem 11.31, Theorem 11.30 and Theorem 11.29 *86*

13 Extensions of Smooth Small Distortions [41]: Introduction *89*
13.1 Class of Sets E *89*
13.2 Main Result *89*

14 Extensions of Smooth Small Distortions: First Results *91*
Lemma 14.1 *91*
Lemma 14.2 *92*
Lemma 14.3 *92*
Lemma 14.4 *93*
Lemma 14.5 *93*

15 Extensions of Smooth Small Distortions: Cubes, Partitions of Unity, Whitney Machinery *95*
15.1 Cubes *95*
15.2 Partition of Unity *95*
15.3 Regularized Distance *95*

16 Extensions of Smooth Small Distortions: Picking Motions *99*
Lemma 16.1 *99*
Lemma 16.2 *101*

17 Extensions of Smooth Small Distortions: Unity Partitions *103*

18 Extensions of Smooth Small Distortions: Function Extension *105*
Lemma 18.1 *105*
Lemma 18.2 *106*

19 Equidistribution: Extremal Newtonian-like Configurations, Group Invariant Discrepancy, Finite Fields, Combinatorial Designs, Linear Independent Vectors, Matroids and the Maximum Distance Separable Conjecture *109*
19.1 s-extremal Configurations and Newtonian $\boldsymbol{s}$-energy *109*
19.2 $[-1,1]$ *110*
19.2.1 Critical Transition *110*
19.2.2 Distribution of $\boldsymbol{s}$-extremal Configurations *111*
19.2.3 Equally Spaced Points for Interpolation *112*
19.3 The n-dimensional Sphere, S^n Embedded in $\mathbb{R}^{n+1}$ *112*
19.3.1 Critical Transition *112*
19.4 Torus *113*
19.5 Separation Radius and Mesh Norm for s-extremal Configurations *114*
19.5.1 Separation Radius of $s > n$-extremal Configurations on a Set Y^n *116*
19.5.2 Separation Radius of $s < n-1$-extremal Configurations on S^n *116*
19.5.3 Mesh Norm of s-extremal Configurations on a Set Y^n *116*
19.6 Discrepancy of Measures, Group Invariance *117*
19.7 Finite Field Algorithm *119*
19.7.1 Examples *120*
19.7.2 Spherical $\hat{t}$-designs *120*
19.7.3 Extension to Finite Fields of Odd Prime Powers *121*
19.8 Combinatorial Designs, Linearly Independent Vectors, MDS Conjecture *121*
19.8.1 The Case $q = 2$ *122*
19.8.2 The General Case *122*
19.8.3 The Maximum Distance Separable Conjecture *123*

20 Covering of $SU(2)$ and Quantum Lattices *125*
20.1 Structure of $SU(2)$ *126*
20.2 Universal Sets *127*
20.3 Covering Exponent *128*
20.4 An Efficient Universal Set in PSU(2) *128*

21 The Unlabeled Correspondence Configuration Problem and Optimal Transport *131*
21.1 Unlabeled Correspondence Configuration Problem *131*
21.1.1 Non-reconstructible Configurations *131*
21.1.2 Example *132*
21.1.3 Partition Into Polygons *134*
21.1.4 Considering Areas of Triangles—*10-step Algorithm* *134*
21.1.5 Graph Point of View *137*
21.1.6 Considering Areas of Quadrilaterals *137*
21.1.7 Partition Into Polygons for Small Distorted Pairwise Distances *138*
21.1.8 Areas of Triangles for Small Distorted Pairwise Distances *138*
21.1.9 Considering Areas of Triangles (part 2) *141*
21.1.10 Areas of Quadrilaterals for Small Distorted Pairwise Distances *142*
21.1.11 Considering Areas of Quadrilaterals (part 2) *145*

22 A Short Section on Optimal Transport *147*

23 Conclusion *149*

References *151*
Index *159*

Preface

This monograph, for fixed integers $n \geq 1$ and $d \geq 2$, provides a modern treatment of Whitney extensions of near isometries in $\mathbb{R}^n$, non-rigid alignment of data in $\mathbb{R}^d$, interpolation by near isometries in $\mathbb{R}^d$, continuum limits of shortest paths in $\mathbb{R}^d$ with several intersecting topics in pure and applied harmonic analysis and data science for example, approximation of near distortions by elements of the orthogonal group $O(d)$ using the space of functions of bounded mean oscillation (BMO), clustering of data in $\mathbb{R}^d$, equidistribution and discrepancy via minimal energy and finite field techniques, tensor methods in Whitney theory, approximation theory in algebraic geometry, quantum lattices and covers of the special unitary group $SU(2)$, techniques for counting linear independent vectors over finite fields and the maximum distance separable (MDS) conjecture.

The monograph is structured as follows. Chapter 1 introduces the near Whitney extension problem. For finite sets of distinct data in $\mathbb{R}^d$ with varying geometries, the chapter interprets this extension problem as an interpolation, non-rigid alignment problem of distinct data in $\mathbb{R}^d$ and discusses rigid and non-rigid data alignment problems. The monograph, then moves to machinery to analyze the near Whitney extension problem in Chapter 1, from various perspectives for finite sets of distinct points with varying geometries. Chapter 2 introduces the idea of near distorted diffeomorphism extensions which agree with Euclidean motions in $\mathbb{R}^d$. Slow Twists and Slides are introduced as examples of near distorted diffeomorphisms.

Chapter 5 deals with clustering methods of data in $\mathbb{R}^d$ and studies continuum limits of shortest paths. Chapter 7 deals with tensor methods, Chapter 8 studies the use of approximation theory for varieties in algebraic geometry and Chapter 9 introduces near reflection theory.

Chapter 10 studies approximation of near distortions by elements of the orthogonal group $O(d)$, using the space of functions of bounded mean oscillation (BMO) and the John Nirenberg inequality.

Chapters 3, 6, 11–12 introduce Gluing techniques, partitions of unity, further Whitney machinery and finite principles.

Chapter 4 deals with manifold learning and the Johnson–Lindenstrauss theorem.

Chapters 13–18 deal with the analysis of the near Whitney extension problem for compact sets in open sets in $\mathbb{R}^n$. This chapter introduces Whitney techniques such as Whitney cubes and regularization. It provides near distortions agreeing with Euclidean motions in $\mathbb{R}^n$.

Chapter 19 deals with equidistribution and studies minimal energy on n-dimensional compact sets embedded in $\mathbb{R}^{n+1}$ via extremal Newtonian like configurations. It studies group invariant discrepancy, finite field discrepancy, combinatorial designs, techniques for counting linear independent vectors over finite fields, and discusses the maximum distance separable (MDS) conjecture.

Chapter 20 deals with quantum lattices and covers of the special unitary group $SU(2)$.

Finally Chapter 21 deals with the near unlabeled data alignment problem and the related optimal transport problem.

Acknowledgements

The work in this monograph is joint with many collaborators and I thank them for exciting and fruitful collaborations. In particular, I would like to thank my collaborator Charles Fefferman for the collaborative work in our papers [39–42] (Chapters 2–3, 6–7, 9–18, Sections 5.7, 8.1). It is a pleasure to thank many colleagues who have generously supported me with this project. In particular, I would like to mention John Bennedeto, Tony Bloch, Emmanuel Candese, David Ragozin, Kai Diethelm, Nadav Dym, Keaton Hamm, Alfred Hero, Joe Kileel, Victor Lieberman, Roy Liederman, Doron Lubinsky, Daniel McKenzie, Boaz Nadler, Peter Oliver, Peter Sarnak, Michael Sears, Amit Singer, Sungjin Wang and Michael Werman. I thank the referees for the enormous amount of methodical work they undertook checking everything and for their many generous suggestions to improve the monograph. Finally, I wish to thank all the editorial staff at John Wiley & Sons for their support and to acknowledge their expertise in bringing the memoir to its current form.* Given the enormous literature on some of the topics discussed in this monograph, any relevant omissions in our reference listed are unintentional.

*Research support from the National Science Foundation, Georgia Southern University, American Mathematical Society, South African Center for High Performance Computing, and Princeton University is thankfully acknowledged.

This monograph will be of interest to applied and pure mathematicians, computer scientists and engineers working in algebraic geometry, approximation theory, computer vision, data science, differential geometry, harmonic analysis, applied harmonic analysis, manifold and machine learning, networks, optimal transport, partial differential equations, probability, shortest paths, signal processing and neuroscience. I hope that readers will enjoy the book and will think about the many areas open to investigation detailed within. I hope that this monograph will inspire new research and curiosity.

Steven B. Damelin
Ann Arbor, MI
18 August, 2023

Overview

Notation: Throughout, $d \geq 2$ and $n \geq 1$ will be fixed positive integers. By $|.|$, we mean the Euclidean norm on $\mathbb{R}^n$*. A function $A : \mathbb{R}^n \to \mathbb{R}^n$ is an improper Euclidean motion (rigid motion) if there exist $M \in O(n)$ and a translation $x_0 \in \mathbb{R}^n$ so that for every $x \in \mathbb{R}^n$, $A(x) = Mx + x_0$. If $M \in SO(n)$, then A is a proper Euclidean motion. Here, $O(n)$ and $SO(n)$ are respectively the orthogonal and special orthogonal groups. A Euclidean motion can either be proper or improper. We will call a function $f : \mathbb{R}^n \to \mathbb{R}^n$, a c-distortion if there exists $c > 0$ small enough depending on n so that $(1-c)|x-y| \leq |f(x)-f(y)| \leq |x-y|(1+c)$ for every $x, y \in \mathbb{R}^n$. Note that f is non-rigid and distorts distances by factors $1 \pm c$ making it a near isometry (almost preserves distances). Rigid motions A are isometries, that is, they are distance preserving and satisfy $|A(x) - A(y)| = |x - y|$ for every $x, y \in \mathbb{R}^n$. A function $f : \mathbb{R}^n \to \mathbb{R}^n$ is bi-Lipschitz if there exists a constant $C \geq 1$ (not depending on n) so that uniformly for all $x, y \in \mathbb{R}^n$, $\frac{1}{C}|x-y| \leq |f(x)-f(y)| \leq C|x-y|$.

All constants depend on the dimension n unless stated otherwise. $c, c', c'', c_1, \ldots C, C', C_1, \ldots$ are always positive constants which depend on n and possibly other quantities. This will be made clear. $X, X_1, X', \ldots$ are compact subsets of $\mathbb{R}^n$ unless stated otherwise. The symbols $f, f_1, \ldots$ are used for functions. We will sometimes write for a function f, $f(x)$. It will be clear from the context what we mean. The notation for constants, sets, and functions may denote the same or different constant, set and function at any given time. The context will be clear. Before a precise definition, we sometimes, as a convention moving forward, use imprecise words or phrases such as "close", "local", "global", "rough", "smooth", and others. We do this deliberately for motivation and easier reading before the reader needs to absorb a precise definition.**

*Unless indicated otherwise.
**The letters $a, b, c, d \ldots$ are unfortunately commonly used in numbering. It will be clear moving forward if $a, b, c, d \ldots$ are used for a numbering or a constant.

We will sometimes write that a particular compact set X (class of compact sets X) has a certain geometry (has certain geometries). We ask the reader to accept such phrases until the exact geometry (geometries) on the given set X (given class of sets X) is defined precisely. Geometries refer to one of many different geometries to be defined precisely when needed.

When we speak to a constant c being small enough, we mean that c is less than a small controlled positive constant. The diameter of a compact set $X \subset \mathbb{R}^n$ is: $\mathrm{diam}(X) := \sup_{x,y \in X} |x - y|$ and if X is a finite set, $\mathrm{card}(X)$ denotes the cardinality of the set X.

Throughout, we often work with special sets and constants. These then have their own designated symbols, for example the set E, the constants $\varepsilon, \delta, \eta$ and so forth. It will be clear what these sets/constants are, when used. The special constants ε and δ will always be small enough. We do however remind the reader of this often.

Structure

A large part of this monograph studies three variants of the following problem (preliminary versions).

(1) **Variant (1): Whitney extensions of near isometries on finite subsets of $\mathbb{R}^d$ (preliminary version).** Let $E := \{y_1, \ldots y_k\} \subset \mathbb{R}^d$ be a finite set of distinct labeled points and $\phi : E \to \mathbb{R}^d$ a near isometry. ($1 \leq i \leq k$ are called the labels of the points y_i). Firstly, how to decide if there exists a smooth near isometry $\Phi : \mathbb{R}^d \to \mathbb{R}^d$.which extends ϕ (that is Φ agrees with ϕ on E) and agrees with Euclidean motions on $\mathbb{R}^d$. Secondly, how to understand when there exists a Euclidean motion $A : \mathbb{R}^d \to \mathbb{R}^d$ such that $A(E)$ is close to E measured in the Euclidean norm. (bi-Lipschitz functions will typically not extend unless C is close to 1).[§] [39, 40]

(2) **Variant (2): Near isometric alignment and interpolation of labeled data in $\mathbb{R}^d$ (preliminary version).** Let $E \subset \mathbb{R}^d$ be a finite set of distinct labeled data[¶][||] and $\phi : E \to \mathbb{R}^d$ a near isometry.
   (a) Interpolation and alignment: how to decide if there exists a smooth near isometry $\Phi : \mathbb{R}^d \to \mathbb{R}^d$, so that Φ interpolates ϕ and Φ agrees with Euclidean motions on $\mathbb{R}^d$?
   (b) How to understand when there exists a Euclidean motion $A : \mathbb{R}^d \to \mathbb{R}^d$ such that $A(E)$ is close to E measured in the Euclidean metric [39, 40].

(3) **Variant (3): Whitney extensions of smooth near isometries on compacts subsets of open subsets of $\mathbb{R}^n$ (preliminary version).** Let $U \subset R^n$ be open and $E \subset U$ compact. Let $\phi : U \to \mathbb{R}^n$ be a smooth near isometry. How

---

[§]The points y_i and ϕ_i ($1 \leq i \leq k$) are matched label-wise, meaning for example y_1 to ϕ_1, y_2 to $\phi_2 \ldots y_k$ to ϕ_k.
[¶]Except for Chapter 21, by points/data we will now always mean labeled points/data.
[||]Variants (1–2) are identical problems. Variant (2) is Variant (1) written in the terminology of data scientists.

to decide if there exists a smooth near isometry $\Phi : \mathbb{R}^n \to \mathbb{R}^n$ which extends ϕ from a neighborhood of E and agrees with Euclidean motions on $\mathbb{R}^n$ [41].

In addition to Variants (1–3) above, the monograph studies the following topics:

(a) Continuum limits of shortest paths and clustering of data in $\mathbb{R}^d$ [70, 95].
(b) Equidistribution and minimal energy on n-dimensional compact sets embedded in $\mathbb{R}^{n+1}$ via extremal Newtonian-like configurations. Group invariant discrepancy, finite field discrepancy, combinatorial designs, techniques for counting linear independent vectors over finite fields and the maximum distance separable (MDS) conjecture [28, 29, 33, 43, 46, 48–51, 54, 55, 109].
(c) Approximation of smooth near distortions by elements of the orthogonal group $O(d)$, using the space of functions of bounded mean oscillation (BMO) and the John Nirenberg inequality [42].
(d) Quantum lattices and covers of the special unitary group $SU(2)$ [74].
(e) Manifold learning and the Johnson–Lindenstrauss theorem.
(f) Unlabeled analogues of variants (1–2) [26].
(g) The optimal transport problem [6].

1

Variants 1–2

In this chapter, we introduce the classical Whitney extension problem. Thereafter, we introduce the near distorted Whitney extension problem and two variants of it. The first, via a purely harmonic analysis problem and the second, translated into a problem related to non-rigid alignment and interpolation of data in $\mathbb{R}^d$. We discuss the Procrustes rigid alignment problem.

1.1 The Whitney Extension Problem

Given a real valued function ϕ on an arbitrary compact set in $\mathbb{R}^n$, the classical Whitney extension problem asks how can one decide whether ϕ extends to a function Φ in $C^m(\mathbb{R}^n)$, $m \geq 1$, the space of real valued functions on $\mathbb{R}^n$ whose derivatives of order m are continuous and bounded? Whitney [114, 115] first studied this problem in 1934. He solved the real line case ($n = 1$) and proved the classic Whitney extension theorem. See [63, 64] and the references cited therein for an interesting account of this problem.

1.2 Variants (1–2) [39, 40]

Problems (1–2) are examples of Variants (1–2).

Problem 1. *Let us be given a positive constant c small enough depending on d. Does there exist a positive constant c′ small enough depending on c so that the following holds? Given two sets of $k \geq 1$ distinct points in $\mathbb{R}^d$, $\{y_1, ..., y_k\}$ and $\{z_1, ..., z_k\}$. Suppose for every $1 \leq i, j \leq k$,*

$$(1 + c')^{-1} \leq \frac{|z_i - z_j|}{|y_i - y_j|} \leq (1 + c'). \tag{1.1}$$

Near Extensions and Alignment of Data in $\mathbb{R}^n$: Whitney extensions of near isometries, shortest paths, equidistribution, clustering and non-rigid alignment of data in Euclidean space,
First Edition. Steven B. Damelin.

(1) *Does there exist a c-distortion* $\Phi : \mathbb{R}^d \to \mathbb{R}^d$ *which obeys* $\Phi(y_i) = z_i$, $1 \le i \le k$?
(2) *Is it possible that* Φ *at the same time agrees with Euclidean motions as well?*
(3) *Can one say something about how* c, c', k, d *are related?*

Problem 2. *Let us be given a positive constant c small enough depending on d. Does there exist a positive constant c′ small enough depending on c so that the following holds? Given two sets of* $k \ge 1$ *distinct labeled points in* $\mathbb{R}^d$, $\{y_1, ..., y_k\}$ *and* $\{z_1, ..., z_k\}$. *Suppose for every* $1 \le i, j \le k$, *(1.1) holds.*

(1) *Is it possible to find a Euclidean motion A which obeys* $A(y_i)$ *is close to* z_i *for every* $1 \le i \le k$. *Here close depends on c and the points* $\{y_1, ..., y_k\}$ *and is measured in the Euclidean norm.*
(2) *Can one say something about how* c, c', k, d *are related?*

Remark 1. *A central remark, at this juncture, is needed moving forward. Problem 1 and Problem 2 are fundamentally different in the sense that Problem 1 is a problem dealing with the existence of extensions. Problem 2 does not ask for an extension. It asks for a Euclidean motion only. This fact translates itself in many ways, for example in how the constants* c, c', k, d *relate to each other.*

In the case of isometry, Remark 1 is far less subtle: indeed, the following result is well-known, see for example [112].

Let $\{y_1, ..., y_k\}$ and $\{z_1, ..., z_k\}$ be two collections of $k \ge 1$ distinct points in $\mathbb{R}^d$. Suppose that the pairwise distances between the points are equal, that is, the two sets of points are isometric. That is

$$|z_i - z_j| = |y_i - y_j|, \ 1 \le i, j \le k.$$

Then, there exists a Euclidean motion, $A : \mathbb{R}^d \to \mathbb{R}^d$ with

$$A(y_i) = z_i, \ 1 \le i \le k.$$

1.3 Variant 2

In this section, we provide some perspective on Variant 2. We briefly discuss the interpolation problem in the sense of manifold learning in Chapter 4.

1.4 Visual Object Recognition and an Equivalence Problem in $\mathbb{R}^d$

Visual object recognition is the ability to perceive properties (such as shape, color and texture) of a visual object in $\mathbb{R}^d$ and to apply semantic attributes to it (such as identifying the visual object). This process includes the understanding of the visual object's use, previous experience with the visual object, and how it relates to the containing space $\mathbb{R}^d$. Regardless of the object's position or illumination, the ability to effectively identify an object, makes the object a "visual" object.

One significant aspect of visual object recognition is the ability to recognize a visual object across varying viewing conditions. These varying conditions include object orientation, lighting, object variability, for example, size, color, and other within-category differences to name just a few. Visual object recognition includes viewpoint-invariant, viewpoint-dependent and multiple view theories to name just a few examples. Visual information gained from an object is often divided into simple geometric components, then matched with the most similar visual object representation that is stored in its memory to provide the object's identification. See the following references and the many cited therein [1, 2, 4–6, 19, 39–42, 59–62, 66, 67, 69, 75, 82, 85–88, 91–93, 100, 101, 107, 111, 118, 120].

With this in mind, we define what we mean by an equivalence problem in $\mathbb{R}^d$. Imagine we are given two visual objects O and O' in $\mathbb{R}^d$. An equivalence and symmetry function $g: O \to O'$, when well defined, is an element of a group. See [98].

Some examples of vision maps are:

(a) Affine functions: a function $A : \mathbb{R}^d \to \mathbb{R}^d$ is an affine function if there exists a linear transformation $M : \mathbb{R}^d \to \mathbb{R}^d$ and $x_0 \in \mathbb{R}^d$ so that for every $x \in \mathbb{R}^d$, $A(x) = Mx + x_0$. Affine functions preserve area (volume) ratios. If M is invertible (i.e., A is then invertible affine), then A is either proper or improper. If M is not invertible, the function A is neither proper nor improper.
(b) Euclidean motions.
(c) Reflections: a reflection $A : \mathbb{R}^d \to \mathbb{R}^d$ is an isometry with a hyperplane as a set of fixed points.
(d) Similarity functions: a Euclidean motion plus a scaling. Similarity functions preserve length ratios.
(e) Projective motions, $(x, y) \to \left(\frac{ax + by + e}{a_1x + b_1y + e_1}, \frac{a_2x + b_2y + e_2}{a_3x + b_3y + e_3}\right)$ with

$$\det \begin{bmatrix} a & b & e \\ a_1 & b_1 & e_1 \\ a_2 & b_2 & e_2 \\ a_3 & b_3 & e_3 \end{bmatrix} = 1.$$

(f) Camera rotations, projective orthogonal transformation:

$$\begin{bmatrix} a_1 & b_1 & e_1 \\ a_2 & b_2 & e_2 \\ a_3 & b_3 & e_3. \end{bmatrix}$$

$\in SO(3)$.

(g) Motion tracking (video group). $(x, y, t) \to (x + at, y + bt, t)$.

- Here, in (e-g), a, b, e, a_i, b_i, e_i, $1 \leq i \leq 3$ are certain real constants. See for example [98].

1.5 Procrustes: The Rigid Alignment Problem

The recent advances in $d = 3$ data acquisition and the increasing interest in augmented and virtual reality have led to an explosion of volumetric data, as exemplified by point clouds. Point cloud data is prevalent in numerous applications, including robotics, autonomous driving, medical imaging, neuroscience, social science and computer graphics. In many of these applications, the captured point clouds correspond to noisy observations of an object/scene undergoing different deformations. One of the core challenges in these applications is to perform point cloud registration, which refers to finding a transformation that aligns or partially aligns the source and target point sets. At a high level, any point cloud registration algorithm must solve two problems: 1) finding accurate correspondences between the points in the source and target point clouds (implicitly or explicitly), and 2) modeling the deformation to match corresponding source and target points. The existing methods then propose different correspondence estimation algorithms and/or propose novel deformation modeling approaches.

The registration/deformation map (i.e., the transformation) could be rigid, or non-rigid. Most of the existing works in the literature have focused on rigid registration of point clouds, as it is a more prevalent problem in classic computer vision tasks like simultaneous localization and mapping (SLAM). The core innovations in these approaches are often with regards to finding the right correspondences between the points. For instance, the classic iterative closest point (ICP) algorithm relies on nearest-neighbor correspondences as measured via the Euclidean distance between points. This section deals with rigid matching. See [1]. The best way to understand the rigid alignment problem is via Procrustes alignment. The classical rigid Procrustes problem is to find a rigid motion that best aligns two

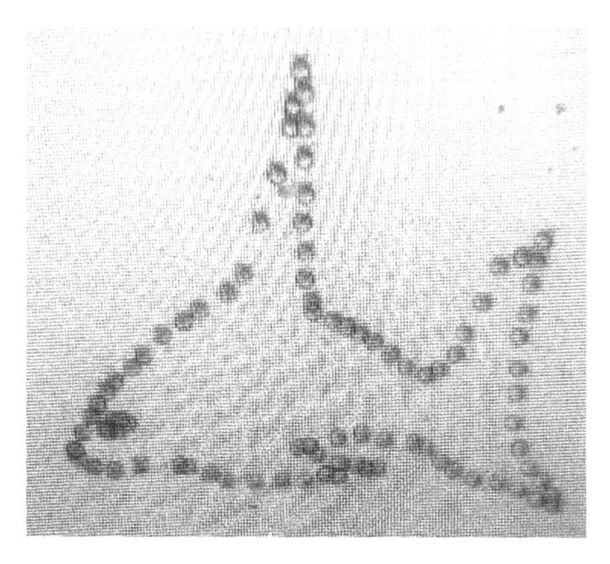

Figure 1.1 Orthogonal Procrustes fish (Rigid): From Simon Ensemble.

given point-sets in the least-squares sense. More precisely, given two sets of $k \geq 1$ distinct points in $\mathbb{R}^d$, say $\{y_1, ..., y_k\}$ and $\{z_1, ..., z_k\}$. The rigid Procrustes problem is the optimization problem

$$\min_{M \in O(d)} \sum_{i=1}^{k} ||M(y_i) - z_i + x_0||_2^2.$$

where $||.||_2$ is the l_2 norm. Here, we recall that $O(d)$ is the orthogonal group of $d \times d$ orthogonal matrices. The alignment is label-wise. Unlabeled problems are challenging, given it is often unclear which point to map to which. See for example our work in Chapter 21.

The rigid Procrustes optimization problem (See Figures 1.1–1.2), has a closed form solution obtained by applying a singular value decomposition (SVD) to a matrix created from the points $\{y_1, ..., y_k\}$ and $\{z_1, ..., z_k\}$. See [105]. For $d = 2, 3$, the rigid Procrustes problem and its generalizations for example to (a) unlabeled problems or/and (b) robustness to outliers, see [1, 59, 61, 62, 93] and the many references cited therein, is a well-known problem with diverse applications for example in computer vision, graphics, robotics [61, 62, 93, 101, 119, 120], chemistry [2], morphology [19], and many others. There are many interesting variants of the rigid Procrustes problem for other norms therein. See also Procrustes analysis with Wasserstein distances, Chapter 22. We do not pursue these directions in this monograph.

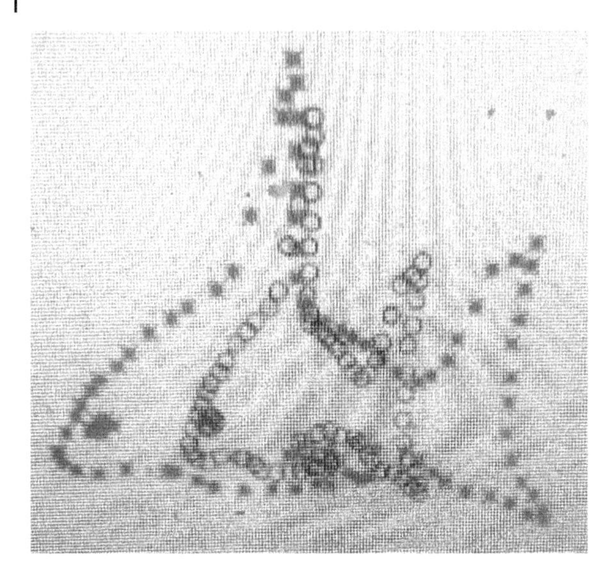

Figure 1.2 Orthogonal Procrustes fish (Rigid): From Simon Ensemble.

1.6 Non-rigid Alignment

In many real-world applications, the deformation between two sets of point cloud data is inherently non-rigid (See Figures 1.3–1.4). For instance, in medical imaging, the point cloud data could come from the surface of a tissue (e.g., liver) or several spiking neurons describing Alzheimer's disease (e.g, in the human brain), both, which can undergo non-rigid deformations. Such nonlinear deformations are generally modeled using two categories of approach, namely parametric and non-parametric approaches. In the parametric approaches, the deformation is characterized via a parametric function, e.g., parameters of an affine transformation or thin plate spline (TPS) parameters, and they are optimized to minimize the expected distance between corresponding source and target points. The non-parametric approaches, on the other hand, directly calculate the displacement (velocity) between source points and their corresponding target points. The existing approaches for non-parametric non-rigid point cloud registration vary in how to estimate the velocity of each point and how to regularize the velocity vector field for coherency and smoothness.

In this section, we rely mostly on two well-known papers: [91, 101]. Non-rigid shape matching often has different challenges to rigid matching even when the space of deformations is limited to, e.g., near isometries and one reason for this is that non-rigid shape matching methods often lead to challenging non-linear, non-convex optimization problems and hence challenging algorithms for robustness. One of the main reasons for this is that unlike in the case of a rigid deformation, non-rigid shape matchings are most frequently represented as pairings

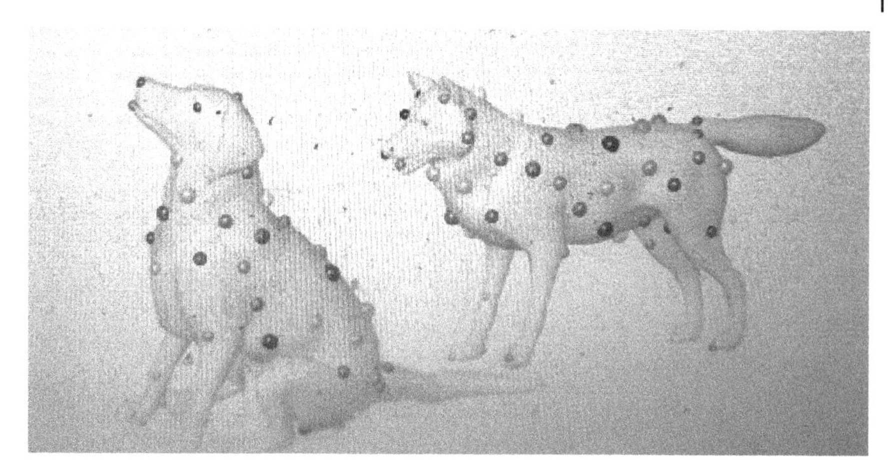

Figure 1.3 Near isometric correspondence shape matching found between a dog and a wolf: see reference [91].

Figure 1.4 Near isometric correspondences found between horses: see reference [101].

(correspondences) of points or regions on the two shapes being matched. Finding correspondences between a given discrete set of points on two different surface grids is an important problem in graphics, geometric processing, and computer vision. For example, applications include shape interpolation, surface completion, statistical shape modeling, symmetry analysis, shape matching, and deformable surface tracking. Intrinsic shapes for objects of the same class are often isometric, and are sometimes composed of big sections that are near isometric. The examples of the wolf, horse, and dog are typical. Exponential sizes of possible point correspondences, support constraints, and shape ambiguity, for example, present challenges to non-rigid matching even in the case of near isometric matching. Landmark matching, function space matching, and matching via Mobius transformations are just three of many interesting ideas known in the study of these difficulties. See [60, 91, 100, 101] and the many authors and papers referenced therein for more details.

Variants 1–2 represent a new idea in the study of near isometric alignment of data. We begin, thus, to build machinery to study Problems 1-2 in Section 2.1.

2

Building ε-distortions: Slow Twists, Slides

In this chapter, we introduce a key object in the near distorted Whitney extension problem, we call a ε-distorted diffeomorphisim which distorts distances by factors $1+\varepsilon$ and $1-\varepsilon$ where $\varepsilon > 0$ is small enough. We introduce two examples of these maps, we call Slow Twists and Slides. We look at their actions and how they can be used to construct ε-distorted diffeomorphisims which agree with Euclidean motions in $\mathbb{R}^d$. [39, 67]

We now begin to look at Problem 1 and Problem 2. We will translate "smooth" into the idea of a c-distorted diffeomorphism (See Figures 2.1–2.12).

2.1 c-distorted Diffeomorphisms

A diffeomorphism $f:\mathbb{R}^d \to \mathbb{R}^d$ is a differentiable, one-to-one and onto function with a differentiable inverse. Given c small enough, a diffeomorphism $f:\mathbb{R}^d \to \mathbb{R}^d$ is a c-distorted diffeomorphism if for every $x, y \in \mathbb{R}^d$, $(1+c)^{-1}I \le (f'(x))^T f'(x) \le (1+c)I$ as matrices. Here, for $d\times d$ matrices, M_1, M_2, M_3, the ordering $M_2 \le M_3 \le M_1$ means as usual that the matrices $M_1 - M_3$ and $M_2 - M_3$ are respectively positive semi-definite and negative semi-definite. Here, also I denotes the identity matrix in $\mathbb{R}^d$.

Moving forward, we will need the following properties of c-distorted diffeomorphisms:

(1) If f is a c-distorted diffeomorphism and $c < c'$, then f is a c'-distorted diffeomorphism.
(2) If f is a c-distorted diffeomorphism then so is f^{-1}.
(3) If f and f_1 are c-distorted diffeomorphisms, then the composition function fof_1 is a $c'c$-distorted diffeomorphism for some constant c'.
(4) If f is a c-distorted diffeomorphism, then f is a c-distortion, that is, $|x-y|(1-c) \le |f(x)-f(y)| \le (1+c)|x-y|$, $x, y \in \mathbb{R}^d$.**

**Since c is small enough we interchange throughout $(1+c)^{-1}$ and $(1-c)$ depending on context.

Near Extensions and Alignment of Data in $\mathbb{R}^n$: Whitney extensions of near isometries, shortest paths, equidistribution, clustering and non-rigid alignment of data in Euclidean space,
First Edition. Steven B. Damelin.

Properties (1–3) follow from the definition. Property (4) follows for example from Bochner's theorem. See for example [112].

We are now going to provide two examples of c-distorted diffeomorphisms.

2.2 Slow Twists

Example 1. *Let $\varepsilon > 0$ and $x \in \mathbb{R}^d$. Let $St(.)$ be the block-diagonal matrix given by*

$$\begin{bmatrix} D_1(x) & 0 & 0 & 0 & 0 & 0 \\ 0 & D_2(x) & 0 & 0 & 0 & 0 \\ 0 & 0 & . & 0 & 0 & 0 \\ 0 & 0 & 0 & . & 0 & 0 \\ 0 & 0 & 0 & 0 & . & 0 \\ 0 & 0 & 0 & 0 & 0 & D_r(x). \end{bmatrix}$$

where for each i either $D_i(x)$ is the 1×1 identity matrix or else

$$D_i(x) = \begin{bmatrix} \cos f_i(|x|) & \sin f_i(|x|) \\ -\sin f_i(|x|) & \cos f_i(|x|) \end{bmatrix}$$

for a smooth function f_i of one variable. Define

$$\Phi_{st}(x) := M^T St(Mx).$$

for $x \in \mathbb{R}^d$ and for a fixed $M \in SO(d)$. One checks that Φ_{st} is ε distorted if

$$A : \; t|f_i'(t)| < c\varepsilon, \; t \geq 0.$$

Here $c > 0$ is small enough. We call Φ_{st} a Slow Twist (in analogy to rotations).

Here are two examples: take a smooth function $f : \mathbb{R} \to \mathbb{R}$ and suppose (A) holds. Then for $x \in \mathbb{R}^2$

$$St(x) = \begin{bmatrix} \cos f(|x|) & \sin f(|x|) \\ -\sin f(|x|) & \cos f(|x|) \end{bmatrix}.$$

Take a smooth function $f : \mathbb{R} \to \mathbb{R}$ and suppose (A) holds. Then for $x \in \mathbb{R}^3$

$$St(x) = \begin{bmatrix} 1 & 0 & 0 \\ 0 & \cos(f(|x|)) & \sin(f(|x|)) \\ 0 & -\sin(f(|x|)) & \cos(f(|x|)) \end{bmatrix}.$$

2.3 Slides

Example 2. *Let $\varepsilon > 0$ and let $f : \mathbb{R}^d \to \mathbb{R}^d$ be a smooth function satisfying the following condition:*

$$B : |f'(t)| < c\varepsilon,\ t \in \mathbb{R}^d.$$

Here, c is small enough. Consider the function $\Phi_{sl}(t) = t + f(t)$. Then Φ_{sl} is ε-distorted and we call it a Slide (in analogy to translations).

The code for Sections (2.4–2.10) below are in [67].

2.4 Slow Twists: Action

Here we illustrate the concept of a Slow Twist on $\mathbb{R}^2$. Given a $\varepsilon > 0$ and a smooth function $f : \mathbb{R} \to \mathbb{R}$ so that (A) holds with the function f. Define the Slow Twist matrix $St(x)$ for any $x \in \mathbb{R}^2$ via

$$St(x) := \begin{bmatrix} \cos f(|x|) & \sin f(|x|) \\ -\sin f(|x|) & \cos f(|x|) \end{bmatrix}.$$

Then given any pure rotation $M \in SO(2)$, the following function $\Phi_{st}(x) := M^T St(Mx) : \mathbb{R}^2 \to \mathbb{R}^2$ is a Slow Twist.

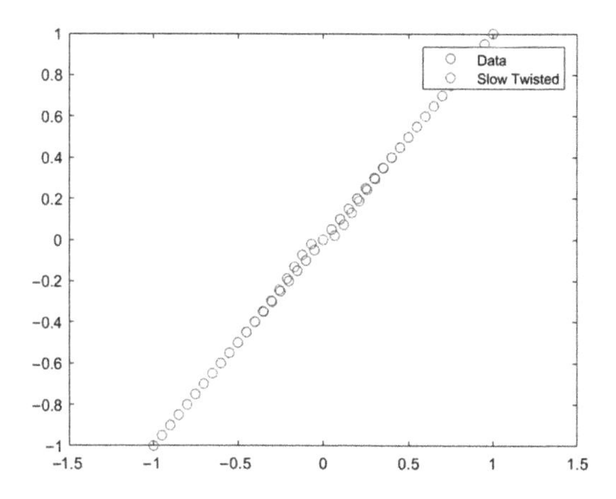

Figure 2.1 Initial points lying on the line $y = x$, and the application of a Slow Twist with $f(x) = \exp(-c|x|)$, with $c = 10$ (top left), $c = 1$ (top middle), and $c = 0.1$ (top right).

(Continued)

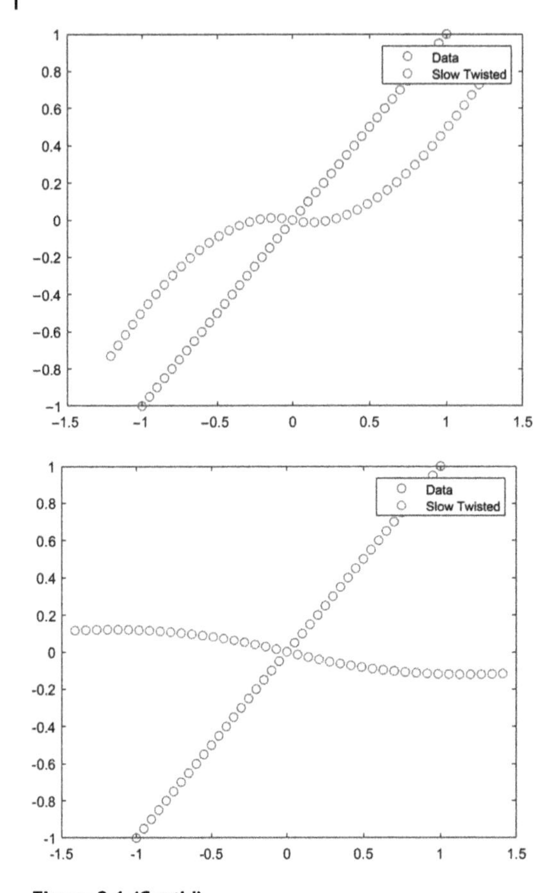

Figure 2.1 (Cont'd)

In two dimensions, the rotation M does not affect anything since rotations are commutative on $\mathbb{R}^2$. However, for higher dimensions this is not the case, and hence we leave them in the formulas, but for now we always fix M to be the identity matrix. For a first set of illustrations, we will look only at one application of a Slow Twist with f being an exponential function with differing scaling parameters.

For large values of c depending on d it can be seen that the twist is near isometric, and even outside a small enough cube centered at the origin, the points are left essentially fixed. On the other hand, as c tends to 0, the twist becomes closer to a pure rotation near the origin. Nevertheless, at a far enough distance, the Slow Twist Φ_{st} will leave the points essentially unchanged.

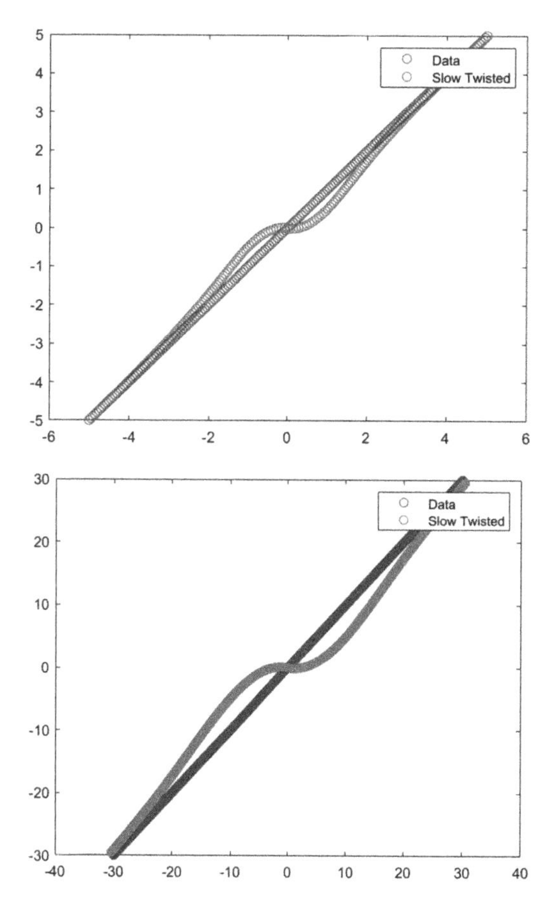

Figure 2.2 Large scale for Slow Twists with $f(x) = \exp(-c|x|)$. Left: $c = 1$, the twist leaves the points essentially static outside $[-5, 5]^2$; Right: $c = 0.1$, the twist only starts to leave the points static outside about $[-30, 30]^2$.

2.5 Fast Twists

Let us pause to consider what happens when the decay condition (A) on the twist function f is not satisfied; in this case we will dub the twist function a Fast Twist for reasons that will become apparent presently.

From Figure 2.3, one can see that when f is the identity function, the rate of twisting is proportional to the distance from the origin, and hence there is no way that the twist function will leave points fixed outside of any ball centered at the origin. Likewise, one sees from Figure 2.4 that the Fast Twist with function $f(x) = |x|^2$ rapidly degenerates points into a jumbled mess.

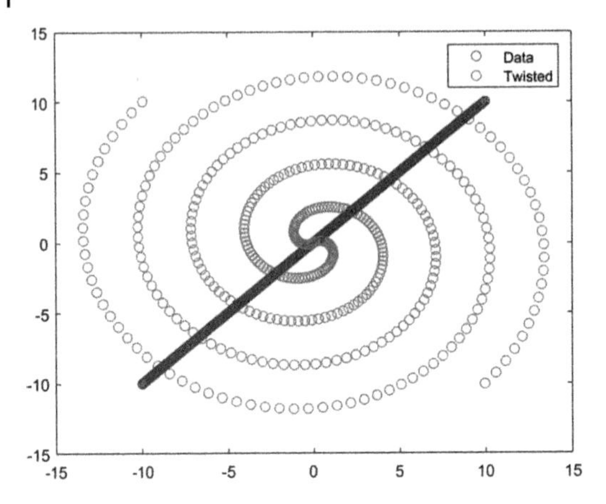

Figure 2.3 Fast Twist with function $f(x) = |x|$.

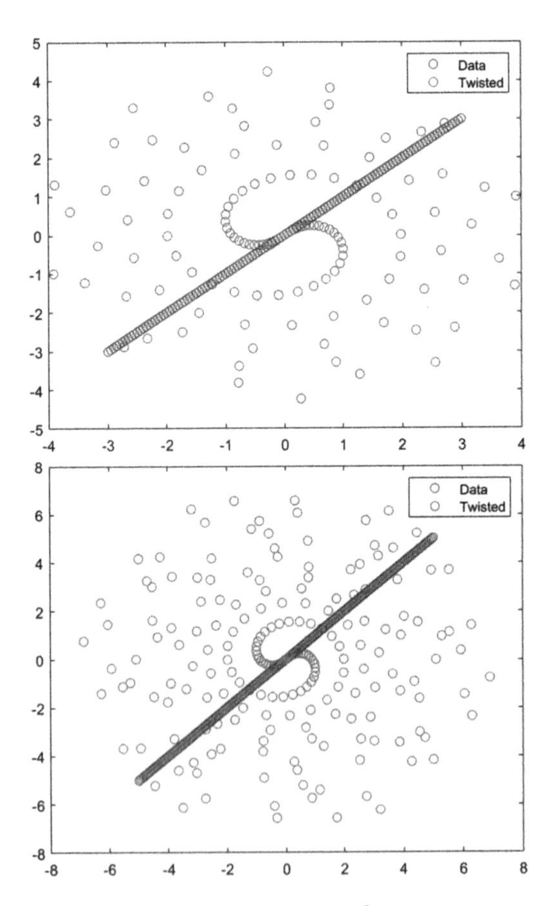

Figure 2.4 Fast Twist with $f(x) = |x|^2$ for a small enough interval $[-3, 3]$ (left) and large interval $[-10, 10]$.

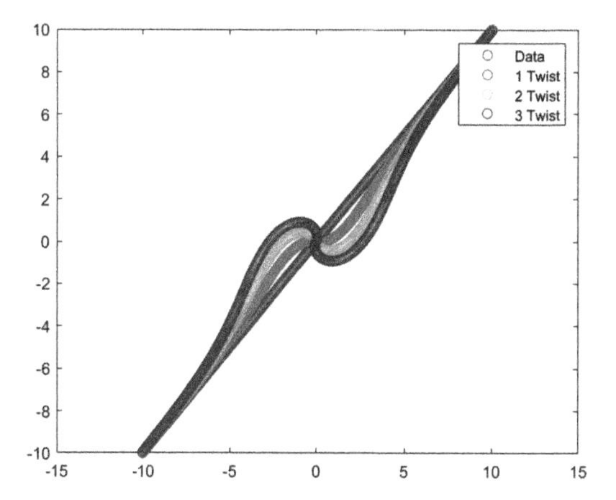

Figure 2.5 Iterated Slow Twist with $f(x) = \exp(-0.5|x|)$. Shown is the initial points along the line $y = x$, $\Phi_{St}(x)$, $\Phi_{st} \circ \Phi_{st}(x)$, and $\Phi_{st} \circ \Phi_{st} \circ \Phi_{st}(x)$.

2.6 Iterated Slow Twists

Here we illustrate what happens when one iteratively applies a Slow Twist to a fixed initial point.

In Figure 2.5, we see an illustration of the fact that the composition of Slow Twists remains a Slow Twist, but the distortion changes slightly; indeed notice that as we take more iterations of the exponential Slow Twist, we have to go farther away from the origin before the new twist leaves the points unchanged.

2.7 Slides: Action

We illustrate some simple examples of Slides on $\mathbb{R}^2$.

First consider equally spaced points on the line $y = -x$, and the Slide given with the function

$$f(t) := \begin{bmatrix} \frac{1}{1+|t_1|^2} \\ \\ \frac{1}{2}e^{-|t_2|} \end{bmatrix}.$$

This is illustrated in Figure 2.6.

To give some more sophisticated examples, we consider first the Slide given with the function

$$f(t) := \begin{bmatrix} e^{-|t_1|} \\ \\ e^{-0.1|t_2|} \end{bmatrix},$$

acting iteratively on uniform points along both the lines $y = x$ and $y = -x$.

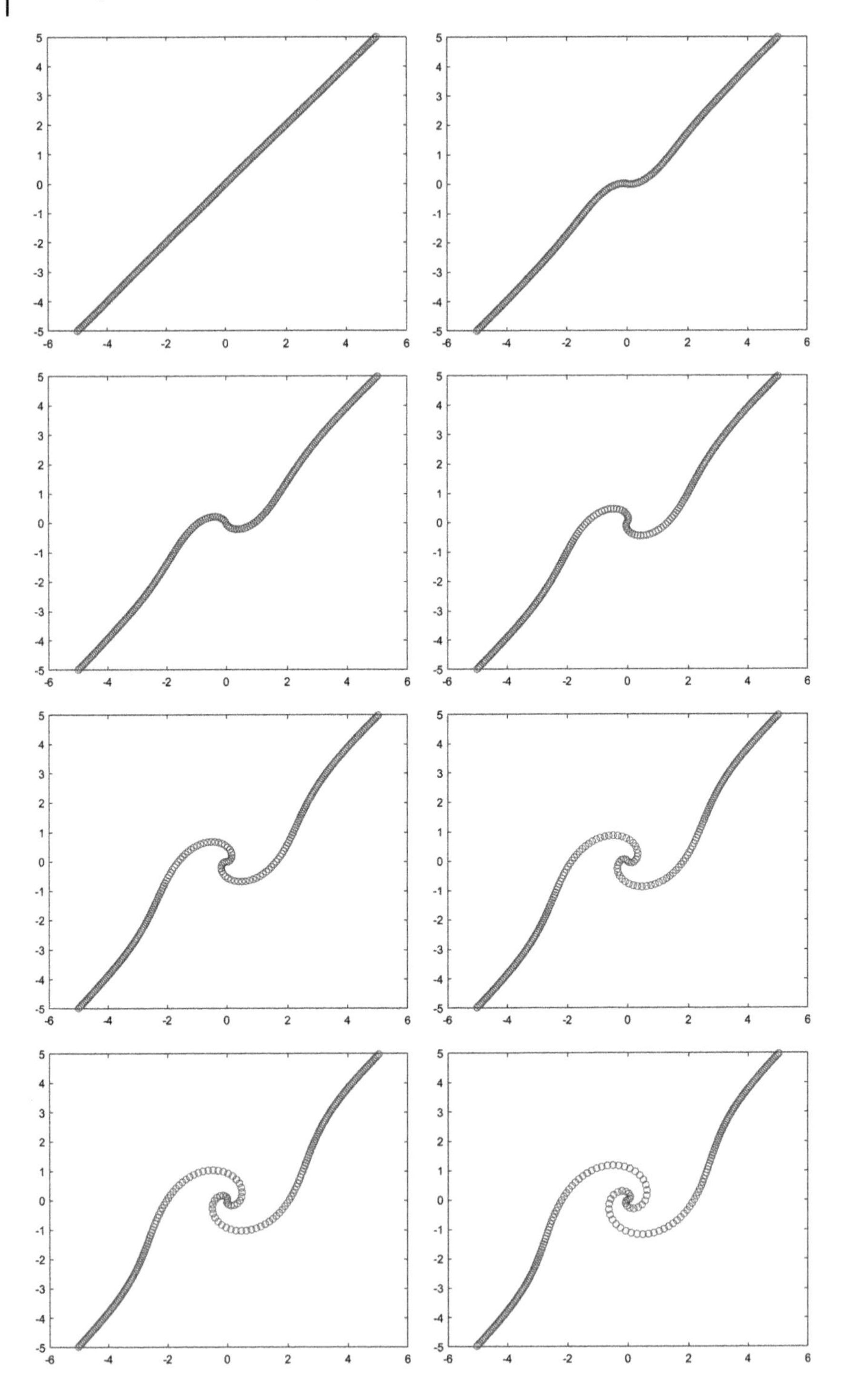
5 4 3 2 1 0 -1 -2 -3 -4 -5
-6 -4 -2 0 2 4 6
5 4 3 2 1 0 -1 -2 -3 -4 -5
-6 -4 -2 0 2 4 6
5 4 3 2 1 0 -1 -2 -3 -4 -5
-6 -4 -2 0 2 4 6
5 4 3 2 1 0 -1 -2 -3 -4 -5
-6 -4 -2 0 2 4 6
5 4 3 2 1 0 -1 -2 -3 -4 -5
-6 -4 -2 0 2 4 6
5 4 3 2 1 0 -1 -2 -3 -4 -5
-6 -4 -2 0 2 4 6
5 4 3 2 1 0 -1 -2 -3 -4 -5
-6 -4 -2 0 2 4 6
5 4 3 2 1 0 -1 -2 -3 -4 -5
-6 -4 -2 0 2 4 6

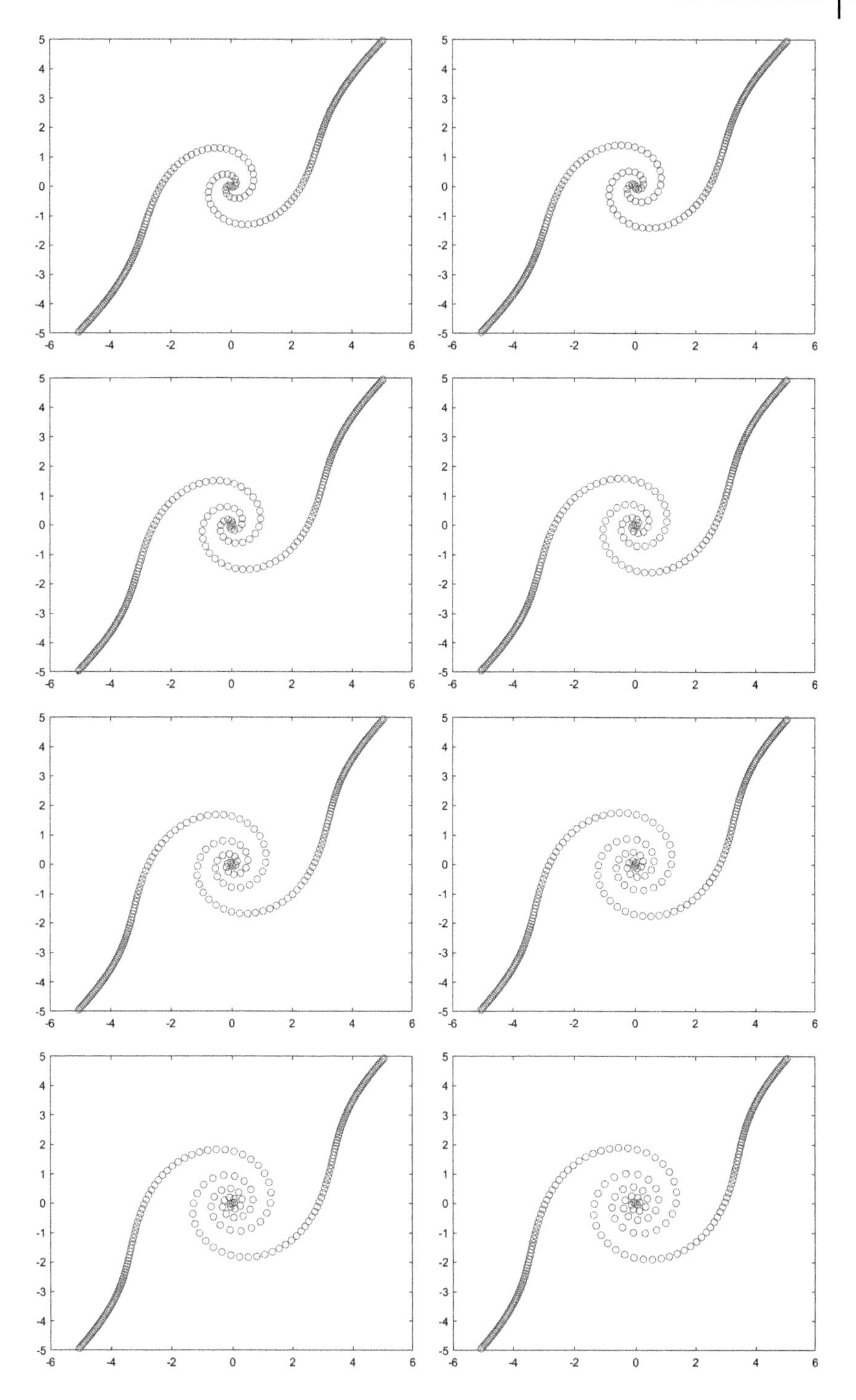
5 4 3 2 1 0 -1 -2 -3 -4 -5
-6 -4 -2 0 2 4 6
5 4 3 2 1 0 -1 -2 -3 -4 -5
-6 -4 -2 0 2 4 6
5 4 3 2 1 0 -1 -2 -3 -4 -5
-6 -4 -2 0 2 4 6
5 4 3 2 1 0 -1 -2 -3 -4 -5
-6 -4 -2 0 2 4 6
5 4 3 2 1 0 -1 -2 -3 -4 -5
-6 -4 -2 0 2 4 6
5 4 3 2 1 0 -1 -2 -3 -4 -5
-6 -4 -2 0 2 4 6
5 4 3 2 1 0 -1 -2 -3 -4 -5
-6 -4 -2 0 2 4 6
5 4 3 2 1 0 -1 -2 -3 -4 -5
-6 -4 -2 0 2 4 6

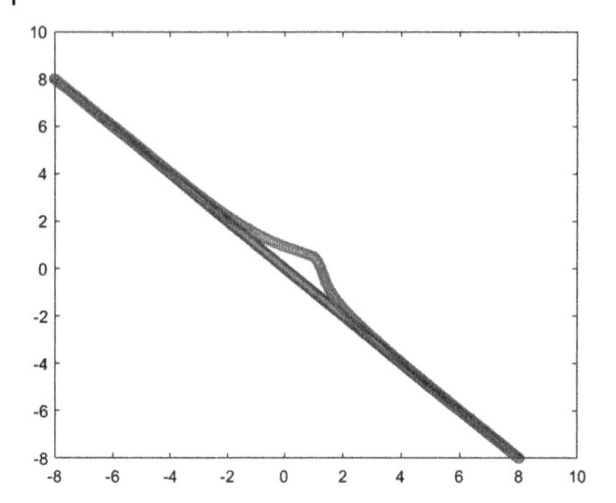

Figure 2.6 Slide with the function f given above.

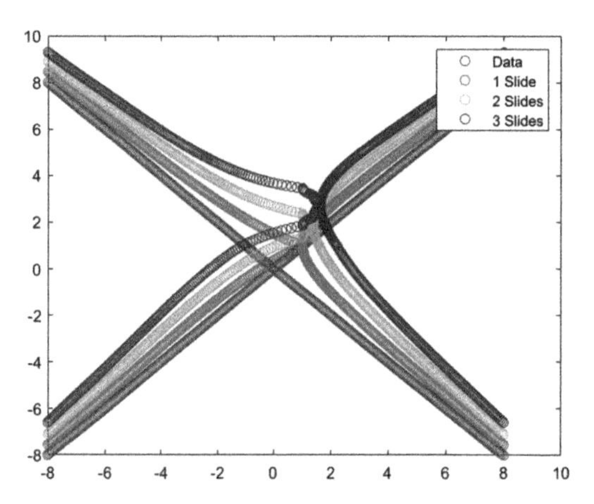

Figure 2.7 Lines $y = x$ and $y = -x$ along with $\Phi_{sl}(x)$, $\Phi_{sl} \circ \Phi_{sl}$ and $\Phi_{sl} \circ \Phi_{sl} \circ \Phi_{sl}(x)$ for f.

Similarly, the following figure shows the Slide given with the function

$$f_2(t) := \begin{bmatrix} 1 - e^{-|t_1|} \\ \\ 1 - e^{-0.1|t_2|} \end{bmatrix}$$

acting iteratively on uniform points along the lines $y = x$ and $y = -x$.

2.8 Slides at Different Distances

To illustrate the effect of the distance of points from the origin, we illustrate here how Slides affect uniform points on circles of different radii.

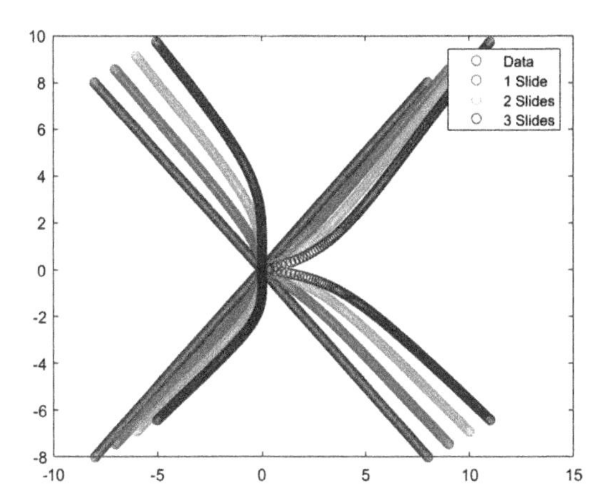

Figure 2.8 Lines $y = x$ and $y = -x$ along with $\Phi_{sl}(x)$, $\Phi_{sl} \circ \Phi_{sl}$ and $\Phi_{sl} \circ \Phi_{sl} \circ \Phi_{sl}(x)$ for f_2.

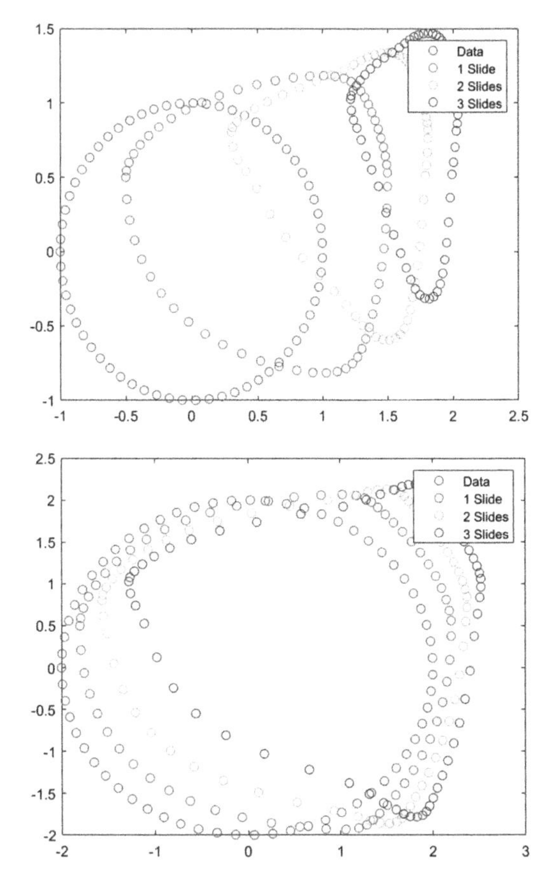

Figure 2.9 Circles under three iterated Slides, beginning with a circle of radius 1 (top left), 2 (top right) and 4 (bottom).

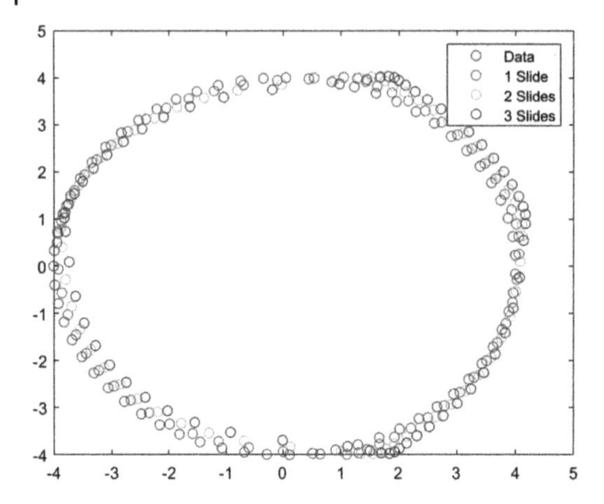

Figure 2.9 (Cont'd)

We use again the asymmetric sliding function

$$f(t) = \begin{bmatrix} \frac{1}{1+|t_1|^2} \\ \\ \frac{1}{2}e^{-|t_2|} \end{bmatrix}.$$

We see from Figure 2.9 that the farther out the points are, i.e., the larger the radius of the initial circle, the less the effect of the Slide, which makes sense given the definition and the fact that the Slides must be ε-distortions of $\mathbb{R}^2$.

2.9 3D Motions

Here we illustrate some of the motions above in $\mathbb{R}^3$.

We construct a generic rotation matrix in $SO(3)$ by specifying parameters a, b, a_1, d satisfying $a^2 + b^2 + a_1^2 + d^2 = 1$, and the rotation matrix M is defined by

$$M = \begin{bmatrix} a^2 + b^2 - a_1^2 - d^2 & 2(ba_1 - ad) & 2(bd + aa_1) \\ 2(ba_1 + ad) & a^2 - b^2 + a_1^2 - d^2 & 2(a_1 d - ab) \\ 2(bd - aa_1) & 2(a_1 d + ab) & a^2 - b^2 - a_1^2 + d^2 \end{bmatrix}.$$

As a reminder, our Slow Twist on $\mathbb{R}^3$ is thus $M^T St(Mx)$.

Example 3. *Our first example is generated by the rotation matrix M as above with parameters* $a = b = \frac{1}{\sqrt{3}}$ *and* $a_1 = d = \frac{1}{\sqrt{6}}$, *and the Slow Twist matrix St as*

$$\begin{bmatrix} 1 & 0 & 0 \\ 0 & \cos(f(|x|)) & \sin(f(|x|)) \\ 0 & -\sin(f(|x|)) & \cos(f(|x|)) \end{bmatrix},$$

where $f(t) = e^{-\frac{t}{2}}$. *Figures 2.10 and 2.11 show two views of the twisted motions generated by these parameters.*

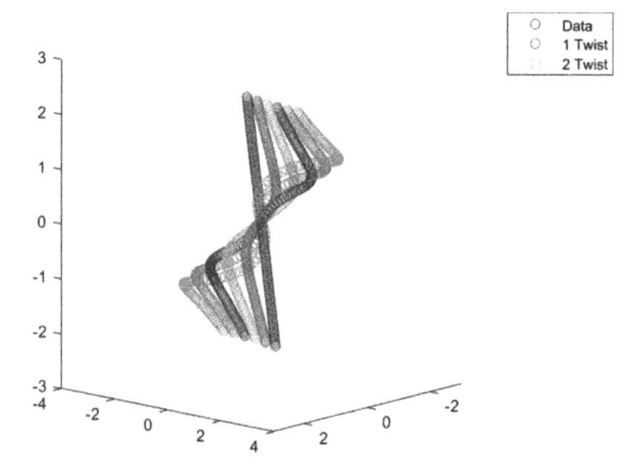

Figure 2.10 Slow Twist in $\mathbb{R}^3$.

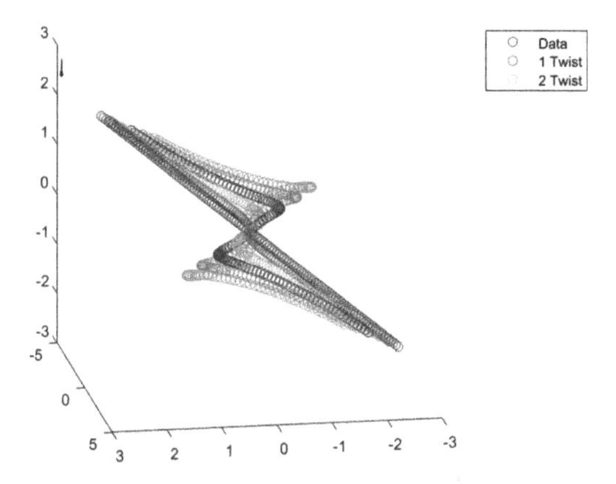

Figure 2.11 Alternate view of Slow Twist in $\mathbb{R}^3$.

2.10 3D Slides

Here we generate 1000 random points on the unit sphere in $\mathbb{R}^3$ and allow them to move under a Slide formed by

$$f(x) = x + \begin{bmatrix} e^{-0.5|x_1|} \\ e^{-|x_2|} \\ e^{-\frac{3}{2}|x_3|} \end{bmatrix}.$$

The results are shown in Figure 2.12.

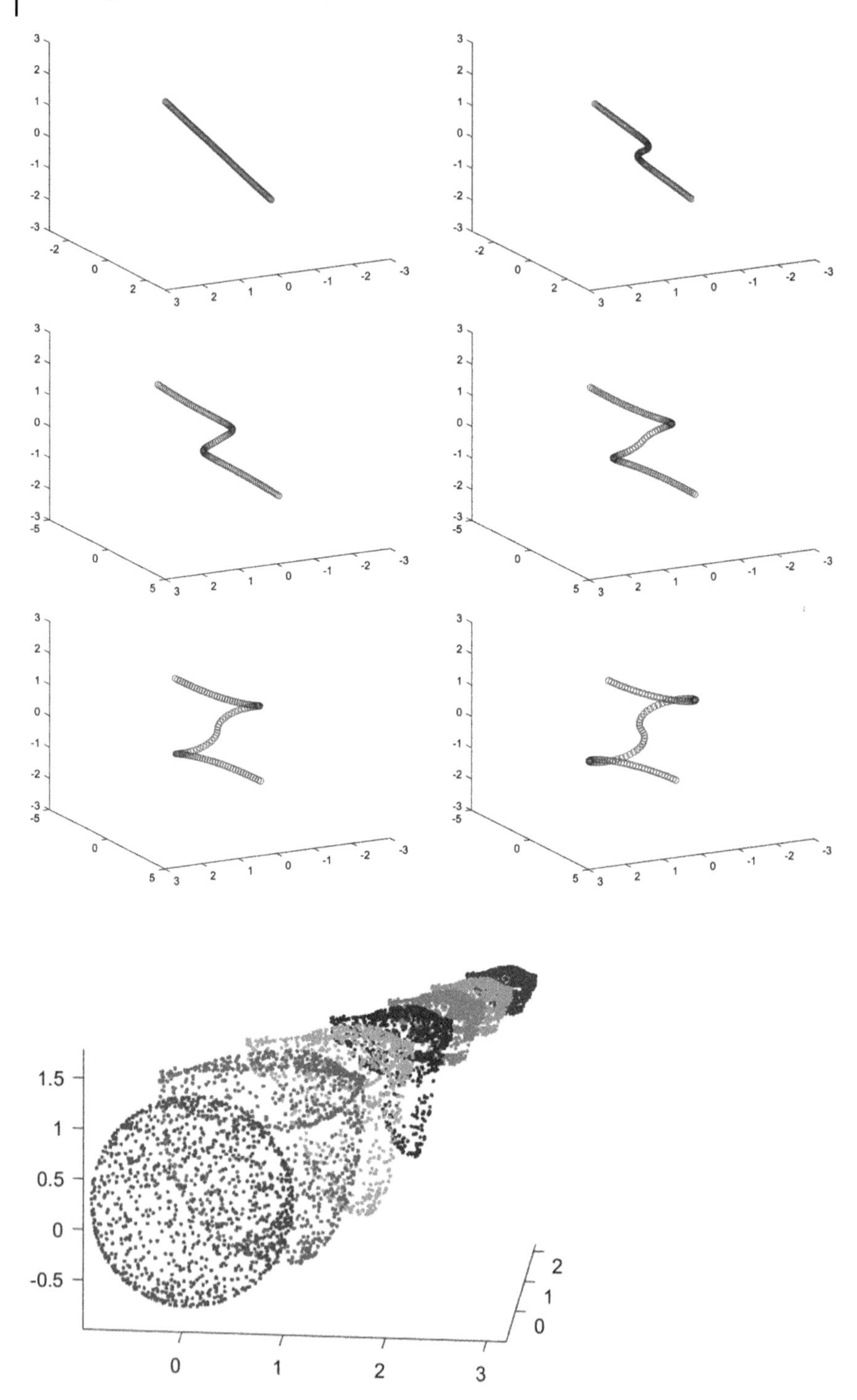

Figure 2.12 Anisotropic Slide on the 2-sphere.

2.11 Slow Twists and Slides: Theorem 2.1

From the definition of Slow Twists and Slides, the following holds [39].

Theorem 2.1. *(1) Let $\varepsilon > 0$. There exists δ_1 small enough depending on ε so that the following holds. Let $M \in SO(d)$ and let $c_1 < \delta_1 c_2$. Then there exists a ε-distorted diffeomorphism f with $f(x) = M(x)$, $|x| \le c_1$ and $f(x) = x$, $|x| \ge c_2$.*

(2) Let $\varepsilon > 0$. There exists δ_1 small enough depending on ε so that the following holds. Let $A(x) := M(x) + x_0$ be a proper Euclidean motion and let $c_3 < \delta_1 c_4$, $|x_0| \le c_5 \varepsilon c_3$. Then there exists a ε-distorted diffeomorphism f with $f(x) = A(x)$, $|x| \le c_3$ and $f(x) = x$, $|x| \ge c_4$.

(3) Let $\varepsilon > 0$. There exists δ_1 small enough depending on ε such that the following holds. Let $c_6 \le \delta_1 c_7$ and let $x, x' \in \mathbb{R}^d$ with $|x - x'| \le c_8 \varepsilon c_6$ and $|x| \le c_6$. Then, there exists a ε-distorted diffeomorphism f with:

(a) $f(x) = x'$.

(b) $f(y) = y$, $|y| \ge c_7$.

Remark 2. *(1) Theorem 2.1 (part(1)) follows from the definition of a Slow Twist.*

(2) Theorem 2.1 (part(2)) follows from the definition of a Slide.

(3) Theorem 2.1 (part(3)) follows from Theorem 2.1 (part(1)) and Theorem 2.1 (part(2)).

(4) Theorem 2.1 exhibits the effect of global rigidity (agreement with the identity far out from the center of mass) and local rigidity.

(5) Procrustes optimization problems for Slow Twists and Slides are challenging with numerous applications in learning, computer vision, and signal processing. For example in remote sensing or photometry, this problem is the well-known coregistration problem in multiple camera hyperspectral data sets.

2.12 Theorem 2.2

We are now ready for our first main result: Theorem 2.2 [39].

Theorem 2.2. *Let $\varepsilon > 0$. Then there exist δ and $\hat{\delta}$ depending on ε small enough such that the following holds. Let $E \subset \mathbb{R}^d$ be a collection of distinct $k \ge 1$ points $E := \{y_1, \ldots, y_k\}$. Suppose we are given a function $\phi : E \to \mathbb{R}^d$ with*

$$|x - y||(1 + \delta)^{-1} \le |\phi(x) - \phi(y)| \le (1 + \delta)|x - y|, \ x, y \in E. \quad (2.1)$$

(1) If $k \le d$, there exists a ε-distorted diffeomorphism $\Phi : \mathbb{R}^d \to \mathbb{R}^d$ so that:

(a) Φ agrees with ϕ on E.

(b) Suppose $y_{i_0} = \phi(y_{i_0})$ for one $i = i_0$, $1 \le i \le k$. Then $\Phi(x) = x$, $|x - y_{i_0}| \ge \hat{\delta}^{-1} \mathrm{diam}\{y_1, \ldots, y_k\}, x \in \mathbb{R}^d$.

(2) *There exists δ_1 such that the following holds. Let $E \subset \mathbb{R}^d$ be a collection of distinct $k \geq 1$ points $E := \{y_1, ..., y_k\}$. Suppose that (2.1) holds with δ_1. There exists a Euclidean motion A with*

$$|\phi(x) - A(x)| \leq \varepsilon diam(E),\ x \in E. \tag{2.2}$$

If $k \leq d$, then A can be taken as proper.

An important observation in Theorem 2.2 is that if $y_i = \phi(y_i)$ for one $i = i_0$, $1 \leq i_0 \leq k$, the extension Φ agrees with a Euclidean motion away from the set E. What this says is that if the function ϕ has a fixed point at one of the points of E, then the function Φ must be essentially rigid away from the set E.

Theorem 2.2 is immediately equivalent to:

Theorem 2.3. *Let $\varepsilon > 0$. Then there exists δ and $\hat{\delta}$ small enough depending on ε such that the following holds. Let $\{y_1, ..., y_k\}$ and $\{z_1, ..., z_k\}$ be two sets of $k \geq 1$ distinct points of $\mathbb{R}^d$. Suppose*

$$|y_i - y_j|(1+\delta)^{-1} \leq |z_i - z_j| \leq (1+\delta)|y_i - y_j|,\ 1 \leq i, j \leq k. \tag{2.3}$$

(1) *If $k \leq d$, there exists a ε-distorted diffeomorphism $\Phi : \mathbb{R}^d \to \mathbb{R}^d$ so that:*
(a) *$\Phi(y_i) = z_i$ for each $1 \leq i \leq k$.*
(b) *Suppose $z_{i_0} = y_{i_0}$ for one $i = i_0$, $1 \leq i \leq k$. Then $\Phi(x) = x$, $|x - y_{i_0}| \geq \hat{\delta}^{-1}$diam $\{y_1, ..., y_k\}, x \in \mathbb{R}^d$.*
(2) *There exists δ_1 such that the following holds. Let $\{y_1, ..., y_k\}$ and $\{z_1, ..., z_k\}$ be two sets of $k \geq 1$ distinct points of $\mathbb{R}^d$. Suppose (2.3) holds with δ_1. There exists a Euclidean motion A so that for $1 \leq i \leq k$,*

$$|A(y_i) - z_i| \leq \varepsilon diam\,(y_1, ..., y_k). \tag{2.4}$$

If $k \leq d$, then A can be taken as proper.

3

Counterexample to Theorem 2.2 (part (1)) for card (E)> d See [39]

We observe immediately the restriction $k \leq d$ (recall $k = card(E)$ for the existence of the extension Φ in Theorem 2.2 (part (1)). Is this simply a "technical issue"? The answer to this "optimistic" guess is no.

The $k \leq d$ sufficient condition for the existence of the extension Φ turns out to be deeper than merely "sufficient" as a tool. In fact, under the geometry of E given by Theorem 2.2, the extension Φ does not always exist for $k > d$. In this chapter we will provide the required counterexample. In fact, the $k > d$ case under the geometry of the finite set E for Theorem 2.2 (part(1)), seems indeed to create a "barrier" to the existence of the extension Φ.

Observe also that there is no such restriction for the existence of the Euclidean motion A in Theorem 2.2 (part (2)). Indeed, the relationships between the cardinality of the set E and the dimension d have no bearing on Theorem 2.2 (part (2)).

3.1 Theorem 2.2 (part (1)), Counterexample: $k > d$

Let us now look at the counterexample.

We fix $2d+1$ points as follows. Let δ be a small enough positive number depending on d. Let $y_1, ..., y_{d+1} \in \mathbb{R}^d$ be the vertices of a regular simplex, all lying on the sphere of radius δ about the origin. Then define $y_{d+2} ... y_{2d+1} \in \mathbb{R}^d$ such that $y_{d+1}, ..., y_{2d+1}$ are the vertices of a regular simplex, all lying in a sphere of radius 1, centered at some point $x_0 \in \mathbb{R}^d$. Next, we define a function

$$\phi : \{y_1, ..., y_{2d+1}\} \to \{y_1, ..., y_{2d+1}\}$$

as follows. We take $\phi|_{\{y_1,...,y_{d+1}\}}$ to be an odd permutation that fixes y_{d+1}, and take $\phi|_{\{y_{d+1},...,y_{2d+1}\}}$ to be the identity. The function ϕ distorts distances by, at most, a

Near Extensions and Alignment of Data in $\mathbb{R}^n$: Whitney extensions of near isometries, shortest paths, equidistribution, clustering and non-rigid alignment of data in Euclidean space,
First Edition. Steven B. Damelin.

factor $1 + c\delta$. Here, we can take δ arbitrarily small enough. On the other hand, for small enough ε, we will show that ϕ cannot be extended to a function $\Phi : \mathbb{R}^d \to \mathbb{R}^d$ satisfying

$$(1+\varepsilon)^{-1}|x-x'| \le |\Phi(x) - \Phi(x')| \le |x-x'|(1+\varepsilon),\ x, x' \in \mathbb{R}^d.$$

In fact, suppose that such a Φ exists. Then Φ is continuous. Note that there exists $M \in O(d)$ with $det(M) = -1$ such that $\phi(y_i) = My_i$ for $i = 1, ..., d+1$. It will be convenient to parametrize the $d-1$ dimensional sphere S^{d-1} embedded in $\mathbb{R}^d$. So let S_t be the sphere of radius $r_t := \delta \cdot (1-t) + 1 \cdot t$ centered at $t \cdot x_0$ for $t \in [0,1]$ and let S_t' be the sphere of radius r_t centered at $\Phi(t \cdot x_0)$. Also, let Sh_t be the spherical shell

$$\{x \in \mathbb{R}^d : r_t \cdot (1+\varepsilon)^{-1} \le |x - \Phi(t \cdot x_0)| \le r_t \cdot (1+\varepsilon)\}$$

and let $f_t : Sh_t \to S_t'$ be the projection defined by

$$f_t(x) - \Phi(t \cdot x_0) = \frac{x - \Phi(t \cdot x_0)}{|x - \Phi(t \cdot x_0)|} \cdot r_t.$$

Since Φ agrees with ϕ, we know that

$$|\Phi(x) - Mx| \le c\varepsilon\delta,\ |x| = \delta. \tag{3.1}$$

Since Φ agrees with ϕ, we know that

$$|\Phi(x) - x| \le c\varepsilon,\ |x - x_0| = 1. \tag{3.2}$$

Our assumption that Φ is a near isometry shows that

$$\Phi : S_t \to Sh_t,\ 0 \le t \le 1$$

and

$$(f_t)o(\Phi) : S_t \to S_t',\ 0 \le t \le 1. \tag{3.3}$$

We can, therefore, define a one-parameter family of functions $\hat{f}_t$, $t \in [0,1]$ from the unit sphere to itself by setting

$$\hat{f}_t(x) = \frac{(f_t o\Phi)(tx_0 + r_t x) - \Phi(tx_0)}{|(f_t o\Phi)(tx_0 + r_t x) - \Phi(tx_0)|} = \frac{(f_t o\Phi)(tx_0 + r_t x) - \Phi(tx_0)}{r_t}.$$

From (3.1), we see that $\hat{f}_0$ is a small enough perturbation of the function $M : S^{d-1} \to S^{d-1}$ which has degree -1. From (3.2), we see that $\hat{f}_1$ is a small enough perturbation of the identity. Consequently, the following must hold:

- Degree $\hat{f}_t$ is independent of $t \in [0,1]$.
- Degree $\hat{f}_0 = -1$.
- Degree $\hat{f}_1 = +1$.

This gives the required contradiction. □

3.2 Removing the Barrier $k > d$ in Theorem 2.2 (part (1))

Moving forward we are going to devote a lot of time studying how to remove the barrier of $k > d$ in Theorem 2.2 (part (1)). Indeed, studying the counterexample in Section 3.1 carefully, we make the optimistic guess that the following new geometry on the finite set E is needed to circumvent this barrier. Roughly put, as we will indeed show, we are able to remove the degenerate cases in Theorem 2.2 (part (1)) (motivated by the counterexample $k > d$) by assuming roughly the following additional conditions on the E. See our detailed analysis moving forward.

(1) The diameter of the set E is not too large.
(2) The points of the set E cannot be too close to each other.
(3) The points of the set E are close to a hyperplane in $\mathbb{R}^d$.
(4) $k = \text{Card}(E)$ still has to be finite but no longer bounded by d. Instead, roughly speaking, what is required is that on any $d + 1$ of the k points which form vertices of a relatively voluminous simplex, the mapping ϕ is orientation preserving.

Moving forward, we will make (1–4) rigorous.

4

Manifold Learning, Near-isometric Embeddings, Compressed Sensing, Johnson–Lindenstrauss and Some Applications Related to the near Whitney extension problem

Our work related to the topics in this chapter are the following [3, 25, 32, 33, 44, 47, 116, 117](diffusion, hyperspectral imaging, gene clustering, partial differential equations and random matrices), [27, 39–42](extensions), [53](signal processing and compressed sensing), [70, 95](shortest paths and power weighted clustering).

4.1 Manifold and Deep Learning Via c-distorted Diffeomorphisms

One of the main challenges in high dimensional data analysis, networks, artificial intelligence, neuroscience, optimal transport and many other related areas of research is dealing with the exponential growth of the computational and sample complexity of several needed generic inference tasks as a function of dimension, a phenomenon termed "the curse of dimensionality".

One intuition that has been put forward to lessen or even obviate the impact of this curse is a manifold hypothesis that the data tends to lie on or near a low dimensional submanifold of the ambient space. See for example Figure 4.1. Algorithms and analyses that are based on this hypothesis constitute an enormous area of research of learning and deep learning theory known as manifold learning. Deep relationships between the manifold hypothesis and extension problems are now known. See also Chapter 22. It is an interesting problem to study the connections between the manifold hypothesis, the near Whitney extension problem and the optimal transport problem, see Section 22.

Near Extensions and Alignment of Data in $\mathbb{R}^n$: Whitney extensions of near isometries, shortest paths, equidistribution, clustering and non-rigid alignment of data in Euclidean space,
First Edition. Steven B. Damelin.

Classical linear methods for manifold learning include principal component analysis (PCA), linear multidimensional scaling (MDSc) and singular value decomposition. Some classical and more recent manifold learning algorithms include Isomap, local linear embedding, Laplacian eigenfunctions, diffusion maps on metric spaces, local linear embedding-via Hessians, direct reduction of manifolds themselves, shortest paths, constrained dimensionality reduction, deep learning, topological manifold learning, distances manifolds, partial differential equations-inpainting, and many others.

4.2 Near Isometric Embeddings, Compressive Sensing, Johnson–Lindenstrauss and Applications Related to c-distorted Diffeomorphisms

Consider now the following framework question.

The topic of embeddings that preserve all pairwise distances between data points has been studied in depth and much of the work on this topic is now classical.

The following embedding is interesting and obeys the following L_2 relaxed notion of isometry.

Given $d' \geq 2$ with $d' < d$. Find an embedding $f : \mathbb{R}^d \to \mathbb{R}^{d'}$ with the following property: there exists $c > 0$ small enough depending on d, d' with

$$||x - y||_2^2(1 - c) \leq ||f(x) - f(y)||_2^2 \leq (1 + c)||x - y||_2^2, \; x, y \in \mathbb{R}^d.$$

The condition on the function f above is called the "restricted isometry property (RIP)".

Applications of such embeddings f, occur for example in random projection methods for data dimension reduction for example, in signal processing, RIP arises in compressive sensing in the following way, see for example [21, 22, 53]. Compressive sensing is a well-known processing of l sparse signals that can be expressed as the sum of only l elements from a certain basis. An important result in compressive sensing is that if a matrix $M \in \mathbb{R}^{d \times d'}$ satisfies RIP on a certain set X of all l sparse signals, then it is possible to stably recover a sparse signal x from measurements $y - f(x)$, $x \in X$ iff d' is of the order of $l \log(d/l)$.

The existence of the embedding f follows from the "The Johnson Lindenstrauss Lemma" which gives an L_2 relationship between the size of a set of points, to the size of d for a smooth dimension reduction problem. Here is one statement.**

**Other more general statements exist.

Theorem 4.4. *Let $s \in (0,1)$. Let $E \subset \mathbb{R}^d$, a finite set of cardinality $l > 1$. Let $m \geq 1$ satisfy $m = O\left(s^{-2}\log(l)\right)$.*

(1) Then there exists $f : E \to \mathbb{R}^m$ satisfying

$$||x-y||_{2(\mathbb{R}^d)}(1-s) \leq ||f(x)-f(y)||_{2(\mathbb{R}^m)} \leq (1+s)||x-y||_{2(\mathbb{R}^d)}, x \in E,\ y \in \mathbb{R}^d.$$

(2) Suppose in addition we demand that $m = O(d)$ as well as $m = O\left(s^{-2}\log(l)\right)$. Then we obtain a quantitative relation between l and d and the theorem is sharp with $s := \frac{1}{min(d,l)}$.

4.3 Restricted Isometry

Now that we have a good idea what near isometric embeddings may look like, what kinds of invariants do they have? We choose one of interest to us namely "reach" (injectivity radius) which would be interesting to study regarding connections to our work and manifold learning with RIP.

Suppose that $f : \mathbb{R}^d \to \mathbb{R}^{d'}$ is a near isometry in the sense that f satisfies RIP. Here, $d' < d$, $d' \geq 2$.

Reach (Injectivity radius): let $X \subset \mathbb{R}^d$ be a d'-dimensional smooth submanifold embedded in $\mathbb{R}^d$. The reach of X, reach(X) measures how regular the manifold X is and it roughly captures valuable local and global properties of the manifold X. Reach is defined as follows: any point in $\mathbb{R}^d$ within a distance

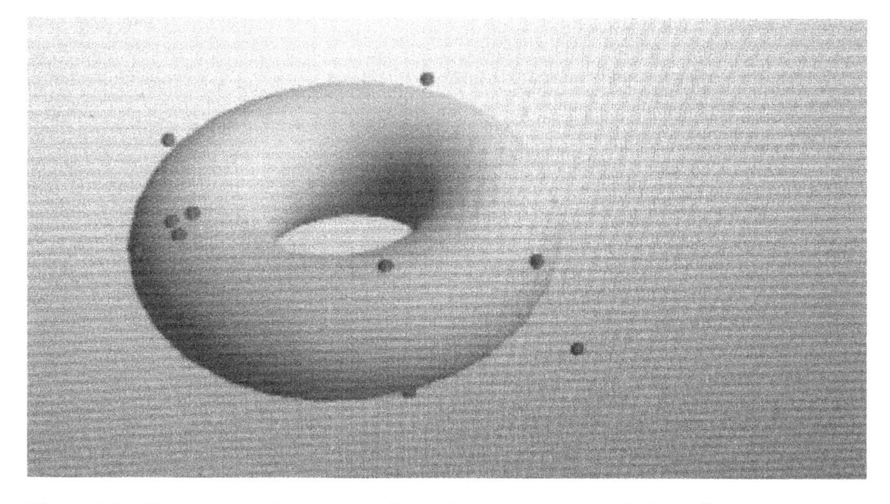

Figure 4.1 Points on or close to a 2-dimensional torus embedded in $\mathbb{R}^3$. See Chapter 19 where we "discretize" a torus by extremal configurations.

reach(X) to X has a unique nearest point on X. For example, the reach of the unit circle is 1 given any point in $\mathbb{R}^2$ whose distance from the unit circle is less than 1 has a unique nearest point on the unit circle. More generally, if X is a $d-1$-dimensional ellipsoid with principal axes $r_1 \geq r_2 \dots \geq r_d > 0$, then reach$(X) = \frac{r_d^2}{r_1}$. Roughly, the reach of X controls how close the manifold X may curve back on itself.

The following holds.

(1) If X is now a smooth, bounded, and boundary-less d'-dimensional submanifold embedded in $\mathbb{R}^d$, then so is $f(X)$.

(2) $\frac{\text{diam}(f(X))}{\text{diam}(X)}$ is close to 1.

(3) $\frac{\text{vol}_{d'}(f(X))}{\text{vol}_{d'}(X)}$ is close to 1.

(4) If f is in addition a rank-d' linear function with its d' non-zero singular values satisfying $[\sigma_{\min}, \sigma_{\max}] \subset (0, \infty)$, then: reach $(f(X)) \geq c\left(\frac{\sigma_{\min}^2}{\sigma_{\max}^3}\right)$(reach($X$)) for some c close enough to 1 depending on d, d'.

(5) Regarding (4), reach $(f(X))$ and reach(X) are typically not close to each other. Suppose the manifold X is concentrated on or close to a d'-dimensional subspace of $\mathbb{R}^d$ and f is an orthogonal projection onto an orthogonal basis for that subspace. These types of orthogonal projections appear in PCA. If d' non-zero singular values of f equal 1 then reach $(f(X))$ is close to reach(X).

(6) Suppose that $f : \mathbb{R}^d \to \mathbb{R}^{d'}$ is an isometry then reach $(f(X)) =$ reach(X)

It is interesting to investigate connections of the work in in Sections (4.2–4.3) to the near Whitney extension problem and the work in Chapter 19 and Chapter 22.

5

Clusters and Partitions

An important tool needed for the proof of Theorem 2.2 (part (1)) and for other results, moving forward, is to "break" up the set E into suitable clusters. This chapter discusses clustering and partitions from various points of view.

5.1 Clusters and Partitions

On an intuitive level, let us suppose that, for example, we are given a n-dimensional compact set X embedded in $\mathbb{R}^{n+1}$ and suppose one requires to produce say 10 000 points which "represent" the set X. How to do this if the set X is described by some geometric property? We may think of a process of "breaking up" a compact set roughly as a "discretization".

Clustering and partitions of sets $X \subset \mathbb{R}^n$ with certain geometry, roughly put, are ways to "discretize" the set X and are used in many mathematical subjects for example harmonic analysis, complex analysis, geometry, approximation theory, data science, number theory, and many more.

When we think of clustering (when possible), we typically speak to finite sets and when we think of partitions (when possible), we typically speak to sets which are not finite.

We provide some examples below with appropriate references. The work [68] is a classic reference for some foundations of the subject of clustering in statistics. Loosely, we think of a clustering of a finite subset $X \subset \mathbb{R}^n$ (when possible), as a finite union of subsets $X_i \subset \mathbb{R}^n$, $i \in I$ for some index set I where roughly "similar" points live in one or a few X_i. We will define "similar" in a moment. Clustering is also affected by the curse of dimensionality. For example, the concept of distance between points in $\mathbb{R}^n$ in a given cluster may easily become distorted as the number of dimensions grows. For example, the distance between any two points in a given cluster may converge in some well-defined sense as the number of dimensions

Near Extensions and Alignment of Data in $\mathbb{R}^n$: Whitney extensions of near isometries, shortest paths, equidistribution, clustering and non-rigid alignment of data in Euclidean space,
First Edition. Steven B. Damelin.

increase. A discrimination then, of the nearest and farthest point in a given cluster can become meaningless.

5.2 Similarity Kernels and Group Invariance

We define, the notion of a "similarity kernel". We take this from our work [33]. Similarity kernels, group invariance, energy, and discrepancy.

Definition 5.5. *Let X^n be a C^∞, compact homogeneous n-dimensional manifold, embedded as the orbit of a compact group G of isometries of $\mathbb{R}^{n'}$, $n' > n$. That is, there exists $x \in X^n$ (a pole) with $G := \{g \cdot x : g \in G\}$. For example, each n-dimensional sphere S^n embedded in $\mathbb{R}^{n+1}$ is the orbit of any unit vector under the action of SO(n+1). A "similarity" kernel $K : X^n \times X^n \to (0, \infty)$ satisfies typically:*

(1) *K is continuous off the diagonal of $X^n \times X^n$ and is lower-semi-continuous on $X^n \times X^n$. Here, the diagonal of $X^n \times X^n$ is the set $\{(x, y) \in X^n \times X^n : x = y\}$.*
(2) *For each $x \in X^n$, $K(x, \cdot)$ and $K(\cdot, y)$ are integrable with respect to surface measure, μ_{Su}, i.e., $K(x, \cdot)$ and $K(\cdot, y)$ are in $L_1(\mu_{Su})$.*
(3) *For each non-trivial finite signed measure $\mu \in X^n$, we have for the energy functional*

$$\int_{X^n}\int_{X^n} K(x, y)d\mu(x)d\mu(y) > 0$$

where the iterated integral may be infinite. This says that K is strictly positive definite.
(4) *$x, y \in X^n$ are "similar" if $K(x, y)$ is small enough.*

A well-known example of a "similarity" kernel involves the Gaussian kernel $K(x, y) := \exp(-c|x - y|)$, $x, y \in \mathbb{R}^n$ where c is scaled data wise. It is an example of a kernel heavily used in non-linear dimension reduction with diffusion functions. We have used this kernel in our work, for example, [24, 25, 33, 44, 95, 116, 117] (Gaussian, dimension reduction-diffusion, hyperspectral image processing, neural net learning, discrepancy and shortest path clustering). Consider the Newtonian: $f(x, y) := f_1(x, y)|x - y|^{-s}$, $0 < s < n$, $x, y \in X^n, x \neq y$. Here, X^n is a certain rectifiable n-dimensional compact set embedded in $\mathbb{R}^{n'}$, $n < n'$ and $f_1 : X^n \times X^n \to (0, \infty)$ is chosen so that f is a "similarity" kernel. For example, X^n can be taken as the d-dimensional sphere, S^n embedded in $\mathbb{R}^{n+1}$ arising from an orbit of a unit vector in the group $SO(n + 1)$. Such kernels are used for example to study extremal configurations on certain n-dimensional compact sets embedded in $\mathbb{R}^{n+1}$ (for example S^n) which form good partitions of these sets useful in many applications in several subjects, for example, approximation theory. See Chapter 19 for more details. See our papers [28, 29, 43, 49–51] (energy, g-invariance, discrepancy).

5.3 Continuum Limits of Shortest Paths Through Random Points and Shortest Path Clustering

In this section, we give a short description of our work in [70, 95] (continuum limits of shortest paths through random points and power weight clustering using shortest paths) on continuum limits of shortest paths through random points and clustering using shortest paths. See Figures 5.1–5.2. We leave the details to our papers [70, 95].

5.3.1 Continuum Limits of Shortest Paths Through Random Points: The Observation

Let $X := \{X_1, X_2, ..., X_k \in \mathbb{R}^d\}$ be i.i.d random vectors with marginal pdf f having compact support S with metric tensor g. Fix two points x_i and x_f in $\mathbb{R}^d$. Let $\gamma = p > 0$ and define G as the complete graph spanning X with weights $\{|X_j - X_i|^p,\ 1 \leq i, j \leq k,\ i \neq j\}$ (the power distance weight). Let $L_p(x_i, x_f)$ be the shortest path between x_i and x_f.

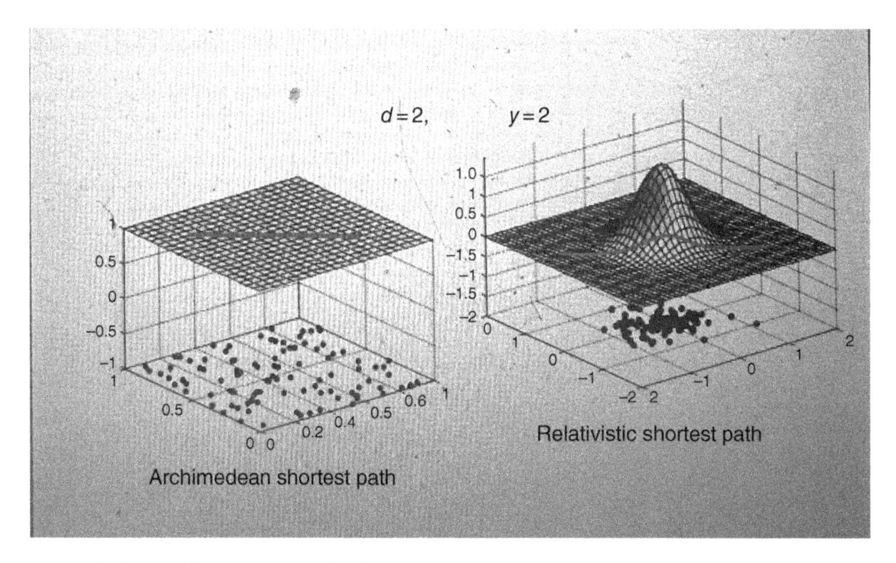

Figure 5.1 Archimean versus the Newtonian.

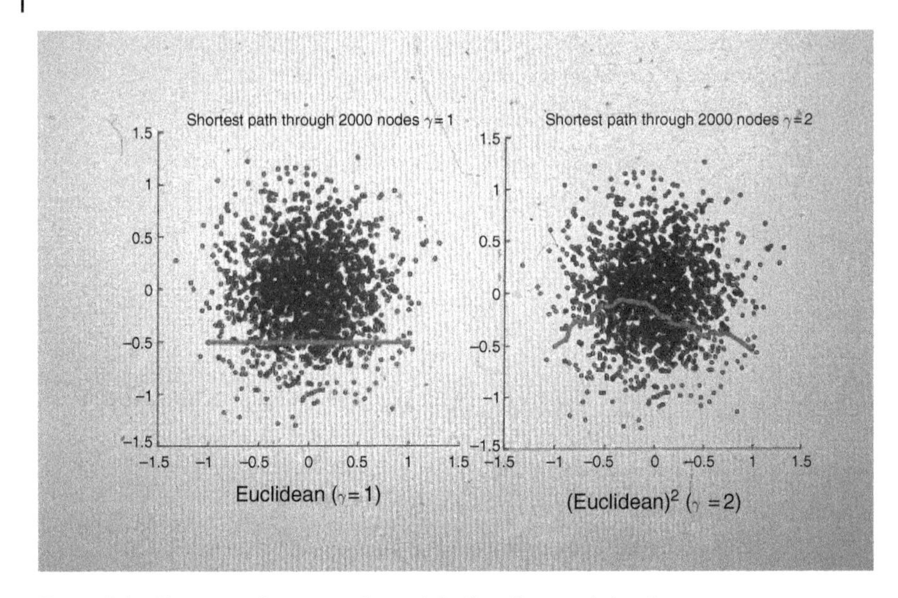

Figure 5.2 For $p \leq 1$: shortest path: straight line. For $p > 1$: lensing.

When $p = 1$, we observe that the shortest path is the Euclidean distance and a straight line. When $p < 1$, the shortest path is still a straight line. Looking at the transfer from uniform distribution to non-uniform distribution, when $p > 1$ the shortest path is no longer a straight line and tends to move to the centre of the distribution favoring denser points. This is analogous to Newtonian–Archimedian versus special relativity where a photon bends at a region of high mass. Thus there is a "lensing effect". How can we explain this as $k \to \infty$?

5.3.2 Continuum Limits of Shortest Paths Through Random Points: The Set Up

We let (M, g_1) be a smooth compact d-dimensional Riemannian manifold without boundary. Consider a probability distribution Pr over Borel subsets of M. Assume that the distribution Pr has a smooth Lebesgue probability density function f with respect to g_1. Let $X_1, X_2, ... X_k$ denote an i.i.d sequence drawn from this density. Now for $p > 1$, called the power parameter, define a new Riemannian metric $g_p = f^{2(1-p)/d} g_1$. That is, if Z_x and W_x are two tangent vectors at a point $x \in M$, then $g_p(Z_x, W_x) = f(x)^{2(1-p)/d} g_1(Z_x, W_x)$. The main result of [70], establishes an asymptotic limit of the lengths of the shortest path through finite subsets of the points $\mathcal{X}_k := \{X_1 .... X_k\}$ as $k \to \infty$. More precisely, if $x, y \in M$, then we let $L_k(x, y)$

denote the shortest path length from x to y through $\mathcal{X}_k \cup \{x, y\}$. Here the edge weight between two points u and v is $\text{dist}_1(u, v)^p$. where dist_1 denotes the Riemannian distance under g_1. The power weighted graph is defined as the complete graph over $\mathcal{X}_k \cup \{x, y\}$. Here, dist_p is then the Riemannian metric under g_p.

5.4 Theorem 5.6

Theorem 5.6. *Suppose that $inf_{x \in M} f(x) > 0$. Then*

$$\lim_{k \to \infty} k^{(1-p)/d} L_k(x, y) = C(d, p)\ dist_p(x, y), c.c$$

with $C(d, p)$, a positive constant independent of f.

We note that more general statements of Theorem 5.6 are given in [70] with rates of convergence. Also note, importantly, that the scaling of the Riemannian metric g_1 by $f^{2(1-p))/d}$ is inversely proportional to the probability density function f. The main result of [70], Theorem 5.6, says that the density f shortens and lengthens paths which respectively pass through regions of high and low density.

5.5 *p*-power Weighted Shortest Path Distance and Longest-leg Path Distance

The work in this section deals with our paper [95] which makes use of our work in [70] on shortest paths through random points drawn from a smooth density supported on a certain set of Riemannian manifolds., c.f the work of [70] in the previous section. For this section, we will keep our discussion short and, as in the previous section, leave the details to our paper [95]. Imagine we have a finite set $X \in \mathbb{R}^d$ we wish to cluster into subsets $X_1, ... X_l$. We assume the manifold hypothesis in the following sense. Assume that X is drawn from a smooth density supported on a certain low dimensional manifold.

Relying on the details in [95], we are going to choose X so that what is below is consistent and well defined. So roughly, let us say x_i and x_j are two different points in X for an appropriate $i, j \in I$, and I an index set. We want to construct clusters of X so that for an appropriate distance the following is true: if x_i and x_j are in the same cluster, the distance between x_i and x_j is small and if x_i and x_j are in different clusters then the distance between x_i and x_j is bounded away from zero. We are going to do this using two distances called the p-weighted shortest path distance and the longest-leg path distance.

Let us look at two kernels suitably scaled.

Given a smooth path $\gamma : x_i \to x_j$, define the $p > 1$-weighted length of γ to be:

$$L^{(p)}(\gamma) := \left(\sum_{i,j} |x_i - x_j|^p \right)^{1/p}.$$

The p-weighted shortest path distance (p-wspm) through X is the minimum weighted length (as above) over all such smooth paths in the sense of

$$K_X^{(p)}(x_i, x_j) := \min \{ L^{(p)}(\gamma) \ : \ \gamma \text{ a smooth path from } x_i \text{ to } x_j \text{ through } X \}.$$

Analogously, given a smooth path $\gamma : x_i \to x_j$, define the longest-leg length of γ as:

$$L^{(\infty)}(\gamma) = \max_{i,j} |x_i - x_j|.$$

The longest-leg path distance (LLPD) through X is the minimum longest-leg length (as above) over all such smooth paths in the sense of

$$K_X^{(\infty)}(x_i, x_j) = \min \{ L^{(\infty)}(\gamma) : \gamma \text{ a smooth path from } x_i \text{ to } x_j \text{ through } X \}.$$

Both kernels $K_{\mathcal{X}}^{(p)}(x_i, x_j)$ and $K_X^{(\infty)}(x_i, x_j)$ have well defined continuous analogues.

The following is proved in [95] for a fixed scaled similarity Gaussian kernel.

- The maximum distance between points in the same cluster is small with high probability, and tends to zero as the number of data points grows without bound. On the other hand, the maximum distance between points in different clusters remains bounded away from zero.
- There exists a modified version of Dijkstra's algorithm that computes k nearest neighbors, with respect to any p-wspm or the LLPD, in $O(k^2 \mathcal{T}_{Enn})$ time, where $\mathcal{T}_{Enn}$ is the cost of a well defined Euclidean nearest-neighbor query.

5.6 *p*-wspm, Well Separation Algorithm Fusion

For a fixed integer $s > 0$, the s-well separation clustering algorithm (s-wsca) [20] is a well-known clustering algorithm whose data structure at a high level is constructed by taking an input node, finding the dimension in which the data spans its largest length and splitting the previously mentioned dimension. The result of the split is two point-sets, in which two new child nodes are formed. The data in each child node is one of the split point sets. Then each newly constructed node undergoes the same process. This is done recursively until the node has a point set which contains a single data point. In our paper [73], we fused the s-wsca algorithm with the p-wspm algorithm from our paper [95] and found that our new algorithm reduces the search space of the p-wspm (Damelin–Mckenzie) algorithm thereby reducing its preprocessing time for certain data sets. We leave the details of this investigation to the paper [73]. Other works of ours on clustering can be found in [44, 74].

5.7 Hierarchical Clustering in $\mathbb{R}^d$ [39]

In this last section in this chapter, we provide the following interesting result which we need, in particular, for the proof of Theorem 2.2 (part (1)). This is hierarchical clustering in $\mathbb{R}^d$ taken from our paper [39].

Theorem 5.7. *Let $k \geq 2$ be a positive integer and let $0 < \eta \leq 1/10$. Let $X \subset \mathbb{R}^d$ be a set consisting of k distinct points. Then, we can partition X into sets $X_1, X_2, \dots, X_{j'_{max}}$ and we can find a positive integer l ($10 \leq l \leq 100 + \binom{k}{2}$) such that the following hold:*

$$diam(X_{j'}) \leq \eta^l \; diam(X), \quad each\ j' \tag{5.1}$$

$$dist(X_{j'}, X_{j''}) \geq \eta^{l-1} \; diam(X), \quad for\ j'' \neq j' \tag{5.2}$$

Proof: We define an equivalence relation $\sim$ on X as follows. $x \sim x'$, for $x, x' \in X$ if and only if $|x - x'| \leq \eta^l$ diam(X) for a fixed positive integer l to be defined in a moment. By the pigeonhole principle, we can always find a positive integer l such that

$$|x - x'| \notin (\eta^l \text{diam}\,(X), \eta^{l-1}\text{diam}\,(X)],\ x, x' \in X.$$

and such that $10 \leq l \leq 100 + \binom{k}{2}$. Let us choose and fix such an l and use it for $\sim$ as defined above. Then $\sim$ is an equivalence relation and the equivalence classes of $\sim$ partition X into the sets $X_1, \dots, X_{j'_{\max}}$ with the properties as required. □

6

The Proof of Theorem 2.3

In this chapter we will establish Theorem 2.3 and hence immediately Theorem 2.2 follows [39].

6.1 Proof of Theorem 2.3 (part(2))

We begin with the proof of Theorem 2.3 (part(2)).

Proof: We argue by way of contradiction that the result does not hold. Then for each $l \geq 1$, we can find points $y_1^{(l)}, ..., y_k^{(l)}$ and $z_1^{(l)}, ..., z_k^{(l)}$ in $\mathbb{R}^d$ satisfying (2.3) with $\delta = 1/l$ but not satisfying (2.4). Without loss of generality, we may suppose that $\mathrm{diam}\left\{y_1^{(l)}, ..., y_k^{(l)}\right\} = 1$ for each l and that $y_1^{(l)} = 0$ and $z_l^{(1)} = 0$ for each l. Thus $|y_i^{(l)}| \leq 1$ for all i and l and

$$(1 + 1/l)^{-1} \leq \frac{|z_i^{(l)} - z_j^{(l)}|}{|y_i^{(l)} - y_j^{(l)}|} \leq (1 + 1/l)$$

for $i \neq j$ and any l. However, for each l, there does not exist a Euclidean motion Φ_0 such that

$$|z_i^{(l)} - \Phi_0(y_i^{(l)})| \leq \varepsilon \tag{6.1}$$

for each i. Passing to a subsequence, $l_1, l_2, l_3, ...,$ we may assume

$$y_i^{(l_\mu)} \to y_i^\infty,\ \mu \to \infty$$

and

$$z_i^{(l_\mu)} \to z_i^\infty,\ \mu \to \infty.$$

Near Extensions and Alignment of Data in $\mathbb{R}^n$: Whitney extensions of near isometries, shortest paths, equidistribution, clustering and non-rigid alignment of data in Euclidean space,
First Edition. Steven B. Damelin.

Here, the points y_i^∞ and z_i^∞ satisfy

$$|z_i^\infty - z_j^\infty| = |y_i^\infty - y_j^\infty|$$

for $i \neq j$. Hence, there is a Euclidean motion $\Phi_0 : \mathbb{R}^d \to \mathbb{R}^d$ such that $\Phi_0(y_i^\infty) = z_i^\infty$. Consequently, for μ large enough, (6.1) holds with l_μ. This contradicts the fact that for each l, there does not exist a Φ_0 satisfying (6.1). Thus, we have proved all the assertions of the theorem except that we can take Φ_0 to be proper if $k \leq d$. To see this, suppose that $k \leq d$ and let Φ_0 be an improper Euclidean motion such that

$$|z_i - \Phi_0(y_i)| \leq \varepsilon \text{diam}\{y_1, ..., y_k\}$$

for each i. There exists an improper Euclidean motion Ψ_0 that fixes $y_1, ..., y_k$ so in place of Φ_0, we may use $\Psi_0 o \Phi_0$ in the conclusion of Theorem 2.3 (part(2)). The proof of Theorem 2.3 (part(2)) is complete. □

Remark 3. *We recall (see Chapter 2), we stated the following:*

(a) *There is no restriction of $k \leq d$ (recall $k = card(E)$) for the existence of the Euclidean motion A in Theorem 2.3 (part (2)) rather the condition $k \leq d$ implies that A can be taken as proper. We see this clearly in the proof of Theorem 2.3 (part (2)).*
(b) *We are going to see in the proof of Theorem 2.3 (part (1)) below that we will need to use Theorem 2.3 (part (2)) and in fact the case when the Euclidean motion A is proper. Then this gives the sufficiency of the restriction $k \leq d$ for the existence of the extension Φ. However, as we have seen from Chapter 3 (the counterexample) this sufficiency condition is not merely something technical which can be removed.*

6.2 A Special Case of the Proof of Theorem 2.3 (part (1))

We now prove a special case of Theorem 2.3 (part (1)). This is given in the following theorem.

Theorem 6.8. *Let $\varepsilon > 0$ and let m be a positive integer. Let $\lambda > 0$ be less than a small enough constant depending only on ε, m and d. Let $\delta > 0$ be less than a small enough constant depending only on λ, ε, m and d. Then the following holds: let $E := y_1, ... y_k$ and $E' := z_1, ... z_k$ be $k \geq 1$ distinct points in $\mathbb{R}^d$ with $k \leq d$ and assume that $y_1 = z_1$. Assume moreover the following:*

$$|y_i - y_j| \geq \lambda^m diam\{y_1, ..., y_k\},\ i \neq j \tag{6.2}$$

and

$$(1+\delta)^{-1} \leq \frac{|z_i - z_j|}{|y_i - y_j|} \leq (1+\delta),\ i \neq j. \tag{6.3}$$

Then, there exists an ε*-distorted diffeomorphism* $\Phi : \mathbb{R}^d \to \mathbb{R}^d$ *such that*

$$\Phi(y_i) = z_i, 1 \leq i \leq k \tag{6.4}$$

and

$$\Phi(x) = x \text{ for } |x - y_1| \geq \lambda^{-1/2} diam\{y_1, ..., y_k\}. \tag{6.5}$$

Proof: Without loss of generality, we may take $y_1 = z_1 = 0$ and $\text{diam}\{y_1, ..., y_k\} = 1$. Applying Theorem 2.3 (part (2)) with $10^{-9}\varepsilon\lambda^{m+5}$ in place of ε, we obtain a proper Euclidean motion

$$A : x \to Mx + x_0 \tag{6.6}$$

such that

$$|\Phi_0(y_i) - z_i| \leq 10^{-9}\varepsilon\lambda^{m+5} \tag{6.7}$$

for each i. In particular, taking $i = 1$ and recalling that $y_1 = z_1 = 0$, we find that

$$|x_0| \leq 10^{-9}\varepsilon\lambda^{m+5}. \tag{6.8}$$

For each i, we consider the balls

$$B_i = B(\Phi_0(y_i), \lambda^{m+3}),\ B_i^+ = B(\Phi_0(y_i), \lambda^{m+1}). \tag{6.9}$$

Note that (6.2) shows that the balls B_i^+ have pairwise disjoint closures since Φ_0 is a Euclidean motion. Applying Theorem 2.1 (part (3)) we obtain for each i, a ε-distorted diffeomorphism $\Psi_i : \mathbb{R}^d \to \mathbb{R}^d$ such that

$$(\Psi_i o \Phi_0)(y_i) = z_i \tag{6.10}$$

and

$$\Psi_i(x) = x \tag{6.11}$$

outside B_i^+. In particular, we see that

$$\Psi_i : B_i^+ \to B_i^+ \tag{6.12}$$

is one-to-one and onto. We may patch the Ψ_i together into a single function $\Psi : \mathbb{R}^d \to \mathbb{R}^d$ by setting

$$\Psi(x) := \left\{ \begin{array}{ll} \Psi_i(x), & x \in B_i^+ \\ x, & x \notin \cup_j B_j^+ \end{array} \right\}. \tag{6.13}$$

Since the B_i^+ have pairwise disjoint closures (6.11) and (6.12) show that Ψ functions $\mathbb{R}^d$ to $\mathbb{R}^d$ and is one-to-one and onto. Moreover, since each Ψ_i is ε-distorted, it now follows easily that

$$\Psi : \mathbb{R}^d \to \mathbb{R}^d \tag{6.14}$$

is an ε-distorted diffeomorphism. From (6.9), (6.10) and (6.13), we also see that

$$(\Psi o \Phi_0)(y_i) = z_i, \forall i. \tag{6.15}$$

Suppose $x \in \mathbb{R}^d$ with $|x| \geq 5$. Then (6.6) and (6.8) show that $|\Phi_0(x)| \geq 4$. On the other hand, each y_i satisfies

$$|y_i| = |y_i - y_1| \leq \text{diam}\,\{y_1, .., y_k\} = 1$$

so another application of (6.6) and (6.8) yields $|\Phi_0(y_i)| \leq 2$. Hence, $\Phi_0(x) \notin B_i^+$, see (6.9). Consequently, (6.13) yields

$$(\Psi o \Phi_0)(x) = \Phi_0(x),\ |x| \geq 5. \tag{6.16}$$

From (6.14), we obtain that

$$\Psi o \Phi : \mathbb{R}^d \to \mathbb{R}^d \tag{6.17}$$

is an ε-distorted diffeomorphism since Φ_0 is a Euclidean motion. Next, applying Theorem 2.1 (part (2)) with $r_1 = 10$ and $r_2 = \lambda^{-1/2}$, we obtain an ε-distorted diffeomorphism $\Psi : \mathbb{R}^d \to \mathbb{R}^d$ such that

$$\Psi_1(x) = \Phi_0(x),\ |x| \leq 10 \tag{6.18}$$

and

$$\Psi_1(x) = x,\ |x| \geq \lambda^{-1/2}. \tag{6.19}$$

Note that Theorem 2.1 (part (2)) applies, thanks to (6.8) and because we may assume $\frac{\lambda^{-1/2}}{10} > \delta_1^{-1}$ with δ_1 as in Theorem 2.1 (part (2)), thanks to our small λ condition.

We now define

$$\tilde{\Psi}(x) := \left\{ \begin{array}{ll} (\Psi o \Phi_0)(x), & |x| \leq 10 \\ \Psi_1(x), & |x| \geq 5 \end{array} \right\}. \tag{6.20}$$

In the overlap region $5 \leq |x| \leq 10$, (6.16) and (6.18) show that $(\Psi o \Phi_0)(x) = \Phi_0(x) = \Psi_1(x)$ so (6.20) makes sense.

We now check that $\tilde{\Psi} : \mathbb{R}^d \to \mathbb{R}^d$ is one-to-one and onto. To do so, we introduce the sphere $S := \{x : |x| = 7\} \subset \mathbb{R}^d$ and partition $\mathbb{R}^d$ into S, inside(S) and outside(S). Since $\Psi_1 : \mathbb{R}^d \to \mathbb{R}^d$ is one-to-one and onto, (6.18) shows that the function

$$\Psi_1 : \text{outside}\,(S) \to \text{outside}\,(\Phi_0(S)) \tag{6.21}$$

is one-to-one and onto. Also, since $\Psi o \Phi_0 : \mathbb{R}^d \to \mathbb{R}^d$ is one-to-one and onto, (6.16) shows that the function

$$\Psi o \Phi_0 : \text{inside}\,(S) \to \text{inside}\,(\Phi_0(S)) \tag{6.22}$$

is one-to-one and onto. In addition, (6.16) shows that the function

$$\Psi o \Phi_0 : (S) \to (\Phi_0(S)) \tag{6.23}$$

is one-to-one and onto. Comparing (6.20) with (6.21), (6.22) and (6.23), we see that $\tilde{\Psi} : \mathbb{R}^d \to \mathbb{R}^d$ is one-to-one and onto. Now since, also $\Psi o \Phi_0$ and Ψ_1 are ε-distorted, it follows at once from (6.20) that $\tilde{\Psi}$ is smooth and

$$(1+\varepsilon)^{-1} I \leq (\tilde{\Psi}'(x))^T \tilde{\Psi}'(x) \leq (1+\varepsilon) I, \; x \in \mathbb{R}^d.$$

Thus,

$$\tilde{\Psi} : \mathbb{R}^d \to \mathbb{R}^d \tag{6.24}$$

is an ε-distorted diffeomorphism. From (6.19), (6.20), we see that $\tilde{\Psi}(x) = x$ for $|x| \geq \lambda^{-1/2}$. From (6.15), (6.20), we have $\tilde{\Psi}(y_i) = z_i$ for each i, since, as we recall,

$$|y_i| = |y_i - y_1| \leq \text{diam}\,\{y_1, \ldots, y_k\} = 1.$$

Thus, $\tilde{\Psi}$ satisfies all the assertions in the statement of the Theorem and the proof of the Theorem is complete. □

6.3 The Remaining Proof of Theorem 2.3 (part (1))

Theorem 2.3 (part (1)) will now follow from the theorem below.

Theorem 6.9. *Given $\varepsilon > 0$, there exists $\lambda, \delta > 0$ depending on ε such that the following holds. Let $E := \{y_1, ..., y_k\}$ and $E' := \{z_1, ..., z_k\}$ be $k \geq 1$ distinct points of $\mathbb{R}^d$ with $1 \leq k \leq d$ and $y_1 = z_1$. Suppose*

$$(1+\delta)^{-1} \leq \frac{|z_i - z_j|}{|y_i - y_j|} \leq (1+\delta),\ i \neq j. \tag{6.25}$$

Then, there exists an ε-distorted diffeomorphism $\Phi : \mathbb{R}^d \to \mathbb{R}^d$ such that

$$\Phi(y_i) = z_i \tag{6.26}$$

for each i and

$$\Phi(x) = x \tag{6.27}$$

for

$$|x - y_1| \geq \lambda^{-1} diam\{y_1, ..., y_k\}.$$

Proof: We use induction on k. For the case $k = 1$, we can take Φ to be the identity function. For the induction step, we fix $k \geq 2$ and suppose we already know the Theorem holds when k is replaced by $k' < k$. We will prove the Theorem holds for the given k. Let $\varepsilon > 0$ be given. We pick small positive numbers $\delta', \lambda_1, \delta$ as follows.

(a) δ' is less than a small enough constant determined by ε, d.
(b) λ_1 is less than a small enough constant determined by δ', d, ε.
(c) δ is less than a small enough constant determined by $\lambda_1, \delta', d, \varepsilon$.

Now let $y_1, ..., y_k, z_1, ..., z_k \in \mathbb{R}^d$ satisfy (6.25). We must produce an ε-distorted diffeomorphism $\Phi : \mathbb{R}^d \to \mathbb{R}^d$ satisfying (6.26) and (6.27) for some λ depending only on $\delta, \lambda_1, \delta', \varepsilon, d$. That will complete the proof of the Theorem.

We apply Theorem 5.7 to $E = \{y_1, ..., y_k\}$ with λ_1 in place of λ. Thus, we obtain an integer l and a partition of E into subsets $E_1, E_2, ..., E_{\mu_{\max}}$ with the following properties:

$$10 \leq l \leq 100 + \binom{k}{2}. \tag{6.28}$$

$$\text{diam}(E_\mu) \leq \lambda_1^l \text{diam}(E) \tag{6.29}$$

for each μ.

$$\text{dist}(E_\mu, E_{\mu'}) \geq \lambda_1^{l-1} \text{diam}(E) \tag{6.30}$$

for $\mu \neq \mu'$. Note that

$$\mathrm{card}(E_\mu) < \mathrm{card}(E) = k \tag{6.31}$$

for each μ thanks to (6.29). For each μ, let

$$I_\mu := \{i : y_i \in E_\mu\}. \tag{6.32}$$

For each μ, we pick a "representative" $i_\mu \in I_\mu$. The $I_1, \dots, I_{\mu_{\max}}$ form a partition of $\{1, \dots, k\}$. Without loss of generality, we may suppose

$$i_1 = 1. \tag{6.33}$$

Define

$$I_{\mathrm{rep}} := \{i_\mu : \mu = 1, \dots, \mu_{\max}\} \tag{6.34}$$

$$E_{\mathrm{rep}} := \left\{y_{i_\mu} : \mu = 1, \dots, \mu_{\max}\right\}. \tag{6.35}$$

From (6.29), (6.30), we obtain

$$(1 - 2\lambda_1^l)\mathrm{diam}(E) \le \mathrm{diam}(E_{\mathrm{rep}}) \le \mathrm{diam}(E),$$

and

$$|x' - x''| \ge \lambda_1^{l-1}\mathrm{diam}(E)$$

for $x, x' \in S_{\mathrm{rep}}$, $x' \ne x''$. Hence,

$$(1/2)\mathrm{diam}(E) \le \mathrm{diam}(E_{\mathrm{rep}}) \le \mathrm{diam}(E) \tag{6.36}$$

and

$$|x' - x''| \ge \lambda_1^m \mathrm{diam}(E_{\mathrm{rep}}) \tag{6.37}$$

for $x', x'' \in E_{\mathrm{rep}}$, $x' \ne x''$ where

$$m = 100 + \binom{d}{2}. \tag{6.38}$$

See (6.28) and recall that $k \le d$. We now apply Theorem 6.8 to the points y_i, $i \in I_{\mathrm{rep}}$, z_i, $i \in I_{\mathrm{rep}}$ with ε in Theorem 6.8 replaced by our present δ'. The hypothesis of Theorem 6.8 holds, thanks to the smallness assumptions on λ_1 and δ. See also (6.38), together with our present hypothesis (6.25). Note also that $1 \in I_{\mathrm{rep}}$ and

$y_1 = z_1$. Thus we obtain a δ'-distorted diffeomorphism $\Phi_0 : \mathbb{R}^d \to \mathbb{R}^d$ such that

$$\Phi_0(y_i) = z_i,\ i \in I_{\text{rep}} \tag{6.39}$$

and

$$\Phi_0(x) = x \text{ for } |x - y_1| \geq \lambda_1^{-1/2} \text{diam}\,\{y_1, \dots, y_k\}. \tag{6.40}$$

Define

$$y_i' = \Phi_0(y_i),\ i = 1, \dots, k. \tag{6.41}$$

Thus,

$$y_{i_\mu}' = z_{i_\mu} \tag{6.42}$$

for each μ and

$$(1 + C\delta')^{-1} \leq \frac{|z_i - z_j|}{|y_i' - y_j'|} \leq (1 + C\delta'),\ i \neq j \tag{6.43}$$

thanks to (6.25), the definition of δ, (6.41) and the fact that $\Phi_0 : \mathbb{R}^d \to \mathbb{R}^d$ is a δ'-distorted diffeomorphism. Now fix $\mu(1 \leq \mu \leq \mu_{\max})$. We now apply our inductive hypothesis with $k' < k$ to the points $y_i', z_i,\ i \in I_\mu$. (Note that the inductive hypothesis applies, thanks to (6.31).) Thus, there exists

$$\lambda_{\text{indhyp}}(d, \varepsilon) > 0,\ \delta_{\text{indhyp}}(d, \varepsilon) > 0 \tag{6.44}$$

such that the following holds: suppose

$$(1 + \delta_{\text{indhyp}})^{-1}|y_i' - y_j'| \leq |z_i - z_j| \leq |y_i' - y_j'|(1 + \delta_{\text{indhyp}}),\ i, j \in I_\mu \tag{6.45}$$

and

$$y_{i_\mu}' = z_{i_\mu}. \tag{6.46}$$

Then there exists a ε distorted diffeomorphism $\Psi_\mu : \mathbb{R}^d \to \mathbb{R}^d$ such that

$$\Psi_\mu(y_i') = z_i,\ i \in I_\mu \tag{6.47}$$

and

$$\Psi_\mu(x) = x,\ \text{for } |x - y_{i_\mu}'| \geq \lambda_{\text{indhyp}}^{-1} \text{diam}(S_\mu). \tag{6.48}$$

We may suppose $C\delta' < \delta_{\text{indhyp}}$ with C as in (6.43), thanks to (6.44) and our smallness assumption on δ'. Similarly, we may suppose that $\lambda_{\text{indhyp}}^{-1} < 1/2\lambda_1^{-1/2}$, thanks to (6.44) and our smallness assumption on λ_1. Thus (6.45) and (6.46) hold, by virtue of (6.43) and (6.42). Hence, for each μ, we obtain an ε-distorted diffeomorphism $\Psi_\mu : \mathbb{R}^d \to \mathbb{R}^d$, satisfying (6.47) and (6.48). In particular, (6.48) yields

$$\Psi_\mu(x) = x, \text{ for } |x - y'_{i_\mu}| \geq 1/2\lambda_1^{-1/2}\text{diam}(E_\mu). \tag{6.49}$$

Taking

$$B_\mu = B(y'_{i_\mu}, 1/2\lambda_1^{-1/2}\text{diam}(E_\mu)), \tag{6.50}$$

we see from (6.49), that

$$\Psi_\mu : B_\mu \to B_\mu \tag{6.51}$$

is one-to-one and onto since Ψ_μ is one-to-one and onto. Next, we note that the balls B_μ are pairwise disjoint.* (Note that the closed ball B_μ is a single point if E_μ is a single point.) This follows from (6.29), (6.30) and the definition (6.50). We may therefore define a function $\Psi : \mathbb{R}^d \to \mathbb{R}^d$ by setting

$$\Psi(x) := \left\{ \begin{array}{ll} \Psi_\mu(x), & x \in B_\mu, \text{ any } \mu \\ x, & x \notin \cup_\mu B_\mu \end{array} \right\}. \tag{6.52}$$

Thanks to (6.51), we see that Ψ is one-to-one and onto. Moreover, since each Ψ_μ is an ε-distorted diffeomorphism satisfying (6.49), we see that Ψ is smooth on $\mathbb{R}^d$ and that

$$(1+\varepsilon)^{-1}I \leq (\Psi'(x))^T\Psi'(x) \leq (1+\varepsilon)I, \ x \in \mathbb{R}^d.$$

Thus,

$$\Psi : \mathbb{R}^d \to \mathbb{R}^d \tag{6.53}$$

is an ε-distorted diffeomorphism. From (6.47) and (6.52), we see that

$$\Psi(y'_i) = z_i, \ i = 1, ..., k. \tag{6.54}$$

Let us define

$$\Phi = \Psi o \Phi_0. \tag{6.55}$$

Thus

$$\Phi \text{ is a } C\varepsilon \text{ -distorted diffeomorphism of } \mathbb{R}^d \to \mathbb{R}^d \tag{6.56}$$

since $\Psi, \Phi_0 : \mathbb{R}^d \to \mathbb{R}^d$ are ε distorted diffeomorphisms. Also

$$\Phi(y_i) = z_i, \ i = 1, ..., k \tag{6.57}$$

as we see from (6.41) and (6.54). Now suppose that

$$|x - y_1| \geq \lambda_1^{-1}\text{diam}\,\{y_1, ..., y_k\}.$$

Since $\Phi_0 : \mathbb{R}^d \to \mathbb{R}^d$ is a ε-distorted diffeomorphism, we have

$$\left|\Phi_0(x) - y_1'\right| \geq (1+\varepsilon)^{-1}\lambda_1^{-1}\text{diam}\{y_1, ..., y_k\} \tag{6.58}$$

and

$$\text{diam}\left\{y_1', ..., y_k'\right\} \leq (1+\varepsilon)\text{diam}\{y_1, ..., y_k\}.$$

See (6.41).

Hence for each μ,

$$\left|\Phi_0(x) - y_{i_\mu}'\right| \geq \left[(1+\varepsilon)^{-1}\lambda_1^{-1} - (1+\varepsilon)\right]\text{diam}\{y_1, ..., y_k\}$$
$$> 1/2\lambda_1^{-1/2}\text{diam}(E_\mu).$$

Thus, $\Phi_0(x) \notin \cup_\mu B_\mu$, see (6.50), and therefore $\Psi o \Phi_0(x) = \Phi_0(x)$, see (6.52). Thus,

$$\Phi(x) = \Phi_0(x). \tag{6.59}$$

From (6.40) and (6.59), we see that $\Phi(x) = x$. Thus, we have shown that

$$|x - y_1| \geq \lambda_1^{-1}\text{diam}\{y_1, ..., y_k\}$$

implies $\Phi(x) = x$. That is, (6.27) holds with $\lambda = \lambda_1$. Since also (6.56) and (6.57) hold we have carried out our inductive step and hence, the proof of the Theorem. □

7

Tensors, Hyperplanes, Near Reflections, Constants (η, τ, K)

In Chapter 3, we presented a counterexample showing that with $k := \text{Card}(E)$, the case $k > d$ in Theorem 2.2 (part(1)) provides a barrier to the existence of the extension Φ there. Moving forward, we now wish to study geometries on the finite set E which remove such a barrier. Indeed, we have already made an optimistic guess for the following geometry on the set E. Roughly, put: we remove degenerative cases as outlined below.

(1) The diameter of the set E is not too large.
(2) The points of the set E cannot be too close to each other.
(3) The points of the set E are close to a hyperplane in $\mathbb{R}^d$.
(4) $k = \text{Card}(E)$ still has to be finite but no longer bounded by d. Instead, roughly speaking, what is required is that on any $d + 1$ of the k points which form vertices of a relatively voluminous simplex, the mapping ϕ is orientation preserving.

In this chapter, we will make (1–4) precise [40].

We mention that, as in the role played by Theorem 2.2 (part(2)) in the proof of Theorem 2.2 (part(1)), in this chapter we are going to establish a variant of Theorem 2.2 (part(2)) which will involve what we call "near reflections".

7.1 Hyperplane; We Meet the Positive Constant η

In this section, we will make mathematically precise what we mean by a finite set say S lying close to a hyperplane in $\mathbb{R}^d$. For this we will use a special constant η.

Definition 7.10. *For a set of $l + 1$ points in $\mathbb{R}^d$, with $l \leq d$, say $z_0, ..., z_l$ we define $V_l(z_0, ..., z_l) := vol_{l \leq d}(simplex_l)$ where $simplex_l$ is the l-simplex with vertices at the*

Near Extensions and Alignment of Data in $\mathbb{R}^n$: Whitney extensions of near isometries, shortest paths, equidistribution, clustering and non-rigid alignment of data in Euclidean space,
First Edition. Steven B. Damelin.

points $z_0, ..., z_l$. Thus $V_l(z_0, ..., z_l)$ is the $l \leq d$-dimensional volume of the l-simplex with vertices at the points $z_0, ..., z_l$.

For a finite set $S \subset \mathbb{R}^d$, we write $V_l(S)$ as the maximum of $V_l(z_0, ..., z_l)$ over all points $z_0, z_1, ..., z_l$ in S. If $V_d(S)$ is small enough, then we expect that S will be close to a hyperplane in $\mathbb{R}^d$. We meet the constant $0 < \eta < 1$, which we will use primarily for an upper bound for $V_d(E)$.

7.2 "Well Separated"; We Meet the Positive Constant τ

It is going to be important that the points of the set E are not too close to each other. We will assume they are "well separated". We meet the constant $0 < \tau < 1$ which we primarily use as a lower bound for the distance $|x - y|$, $x, y \in E$.

7.3 Upper Bound for Card (E); We Meet the Positive Constant K

At the same time we fix the dimension d, we are going to fix a positive constant K which will bound card(E) from above. (K does not depend on d and since E is a finite set, such a K will always exist.)

Our variant of Theorem 2.2 (part (2)) is as follows.

7.4 Theorem 7.11

Theorem 7.11. *Let $0 < \eta < 1$ and $E \subset \mathbb{R}^D$ be a finite set with diam(E) $= 1$. Assume that $V_d(E) \leq \eta^d$. Then, there exists an improper Euclidean motion A and constant $c > 0$ such that*

$$|A(x) - x| \leq c\eta, \; x \in E. \tag{7.1}$$

We now focus on the proof of Theorem 7.11. We need to introduce a near reflection.

7.5 Near Reflections

Suppose that X is a finite subset of a affine hyperplane $X'' \subset \mathbb{R}^d$ with not too large a diameter. Thus X'' has dimension $d - 1$. Let A_1 denote reflection through X''. Then A_1 is an improper Euclidean motion and $A_1(x) = x$ for each $x \in X$. For easy understanding: suppose $d = 2$ and X'' is a line with the set X on the line. Let A_1 denote reflection of the lower half plane to the upper half plane through X. Then A_1 is a Euclidean motion and fixes points on X because it is an isometry. Now Theorem 7.11 constructs an improper Euclidean motion $A(x)$ close to x on the set E where the set E has not too a large diameter and is close enough to a hyperplane. A is called a near reflection.

7.6 Tensors, Wedge Product, and Tensor Product

For the proof of Theorem 7.11, we need to recall some facts about tensors in $\mathbb{R}^n$, wedge product and tensor product.

We recall that $\mathbb{R}^n \otimes \mathbb{R}^{n'}$ is a subspace of $\mathbb{R}^{nn'}$ of dimension nn'. Here, if $e_1, ..., e_n$ and $f_1, ..., f_{n'}$ are the standard basis vectors of $\mathbb{R}^n$ and $\mathbb{R}^{n'}$ respectively, then if $x = (x_1, ..., x_n) \in \mathbb{R}^n$ and $y = (y_1, ..., y_n') \in \mathbb{R}^{n'}$, $x \otimes y$ is a vector in $\mathbb{R}^{nn'}$ spanned by the basis vectors $e_i \otimes f_{i'}$ with coefficients $x_i y_j$. For example, if $n = 2, n' = 3$, then $x \otimes y$ is the vector in $\mathbb{R}^6$ with six components $(x_1y_1, x_1y_2, x_1y_3, y_2x_1, y_2x_2, y_2x_3)$. The wedge product of x, y, $x \wedge y$ is the antisymmetric tensor product $x \otimes y - y \otimes x$. A tensor in $\mathbb{R}^n$ consists of components and also basis vectors associated to each component. The number of components of a tensor in $\mathbb{R}^n$ need not be n. The rank of a tensor in $\mathbb{R}^n$ is the minimum number of basis vectors in $\mathbb{R}^n$ associated to each component of the tensor. ($\mathbb{R}^n$ is realized as the set of all vectors $(p_1..., p_n)$ where each $p_i \in \mathbb{R}$, $1 \leq i \leq n$ are rank-1 tensors given each corresponding component p_i has 1 basis corresponding vector e_i associated to it). A rank-l tensor in $\mathbb{R}^n$ has associated to each of its components l basis vectors out of $e_1, e_2, e_3, ..., e_n$. Real numbers are 0-rank tensors. Now we, let $v_1, ..., v_l \in \mathbb{R}^n$. Writing $v_1 = M_{11}e_1 + ... + M_{l1}e_l, v_2 = M_{12}e_1 + ... + M_{l2}, ..., v_l = M_{1l}e_1 + ... + M_{ll}e_l$ the following holds:

- $v_1 \wedge ... \wedge v_l = \det(M)e_1 \cdot \wedge e_2 ... \wedge e_l$ where M is the matrix

$$\begin{pmatrix} M_{11} & M_{12} & ... & M_{1l} \\ M_{21} & M_{22} & ... & M_{2l} \\ . & . & . & . \\ M_{l1} & M_{l2} & ... & M_{ll} \end{pmatrix}.$$

- $|v_1 \wedge v_2 ... \wedge v_l|^2 = \det(M_1) = \text{vol}_l(v_1, v_2, ..., v_l)$. Here, M_1 is the matrix

$$\begin{pmatrix} v_1.v_1 & v_1.v_2 & ... & v_1.v_l \\ v_2.v_1 & v_2.v_2 & ... & v_2.v_l \\ . & . & . & . \\ v_l.v_1 & v_l.v_2 & ... & v_l.v_l \end{pmatrix}$$

and

$$\mathrm{vol}_l(v_1, v_2, ..., v_l) = \left\{ \sum_{i=1}^{l} c_i v_i \;:\; 0 \leq c_i \leq 1,\; c_i \in \mathbb{R} \right\}$$

is the l-volume of the parallelepiped determined by $v_1, ..., v_l$. $|\cdot|$ is understood here as the rotationally invariant norm on alternating tensors of any rank.

We now prove Theorem 7.11.

Proof: We are going to use tensors and the quantity $V_d(E)$ defined in Definition 7.10.

We have $V_1(E) = 1$ and $V_d(E) \leq \eta^d$. Hence, there exists l with $2 \leq l \leq d$ such that $V_{l-1}(E) > \eta^{l-1}$ but $V_l(E) \leq \eta^l$. Fix such a l. Then there exists a $(l-1)$ simplex with vertices $z_0, ..., z_{l-1} \in E$ and with $(l-1)$-dimensional volume $> \eta^{l-1}$. Fix $z_0, ..., z_{l-1}$. Without loss of generality, we may suppose $z_0 = 0$. Then

$$|z_1 \wedge, ..., \wedge z_{l-1}| > c\eta^{l-1}.$$

yet

$$|z_1 \wedge, ..., \wedge z_{l-1} \wedge x| \leq c'\eta^l,\; x \in E.$$

Now,

$$|z_1 \wedge, ..., \wedge z_{l-1} \wedge x| = |f(x)||z_1 \wedge ... \wedge z_{l-1}|,\; x \in E$$

where f denotes the orthogonal projection from $\mathbb{R}^d$ onto the space of vectors orthogonal to $z_1, ..., z_{l-1}$. Consequently, we have for $x \in E$,

$$c'\eta^l \geq |z_1 \wedge, ..., \wedge z_{l-1} \wedge x| = |f(x)||z_1 \wedge, ..., \wedge z_{l-1}| \geq c\eta^{l-1}|f(x)|.$$

We deduce that we have $|f(x)| \leq c'\eta$ for any $x \in E$. Equivalently, we have shown that every $x \in E$ lies within a distance $c'\eta$ from span$\{z_1, ..., z_{l-1}\}$. This span has dimension $l - 1 \leq d - 1$. Letting H be the hyperplane containing that span and letting A denote the reflection through H, we see that $\mathrm{dist}(x, H) \leq c'\eta$, $x \in E$. Hence,

$$|A(x) - x| \leq c'\eta,\; x \in E.$$

Since A is an improper Euclidean motion, the proof is complete. □

8

Algebraic Geometry: Approximation-varieties, Lojasiewicz, Quantification: (ε, δ)-Theorem 2.2 (part (2))

In this chapter, we are now going to formulate and prove a variant of Theorem 2.2 ((part (2)), namely Theorem 8.12 where we are able to now give quantifications of relations between ε and δ_1 in Theorem 2.2 ((part (2)).

Theorem 8.12 consists of two parts. Part (1) deals with the case when we force our distinct points to lie on an ellipse and in part (2), we assume that our distinct points have not too large diameter and are not too close to each other.

Here is our result.

Theorem 8.12. *The following holds [40]:*

(1) *Let $\delta > 0$ be small enough depending on d. There exist c, c' small enough depending on d such the following holds. Let $\{y_1, ..., y_k\}$ and $\{z_1, ... z_k\}$ be two collections of distinct points in $\mathbb{R}^d$ with*

$$\sum_{i \neq j} |y_i - y_j|^2 + \sum_{i \neq j} |z_i - z_j|^2 = 1, \quad y_1 = z_1 = 0 \tag{8.1}$$

and with diam $\{y_1, ..., y_k\} < 1$, $1 \leq i \leq k$. Suppose that

$$||z_i - z_j| - |y_i - y_j|| < \delta,\ 1 \leq i, j \leq k. \tag{8.2}$$

Then, there exists a Euclidean motion A such that for $1 \leq i \leq k$

$$|z_i - A(y_i)| \leq c\delta^{c'}. \tag{8.3}$$

(2) *Let $0 < \tau < 1$ and fix positive K at the same time as d. There exist c''_K and c'''_K small enough depending on d and K so that the following holds. Let $E := \{y_1, ..., y_k\}$ and $\phi(E)$, $\phi : E \to \mathbb{R}^d$ be two collections of distinct points in $\mathbb{R}^d$ with ϕ satisfying (2.1) for δ_1 satisfying $\delta_1 \leq c''_K \tau^{c'''_K}$. Suppose that diam$(E) \leq 1$,*

Near Extensions and Alignment of Data in $\mathbb{R}^n$: Whitney extensions of near isometries, shortest paths, equidistribution, clustering and non-rigid alignment of data in Euclidean space,
First Edition. Steven B. Damelin.

card(E) $\leq K$ *and* $|x-y| \geq \tau$ *for any* $x, y \in E$ *distinct. Then there is a Euclidean motion A with*

$$|\phi(x) - A(x)| \leq c''_K \delta^{c'''_K},\ x \in E.$$

8.1 Min–max Optimization and Approximation-varieties

Various problems in approximation theory related to smooth varieties have become increasingly popular for various broad interdisciplinary problems, for example, molecule reconstruction and cell morphing, computer vision and shape space. More specifically let us mention the following. We let V be a smooth variety in $\mathbb{R}^d$.

A min–max approximation to V from a point $x \in \mathbb{R}^d$ when it exists is $\mathrm{Inf}_{v \in V} |(x, v)|$. min–max approximants to V from $\mathbb{R}^d$ are unique up to points where $|\cdot|$ is not differentiable and in the latter case the set of points where one does not have uniqueness are nowhere dense in $\mathbb{R}^d$ and lie on some hypersurface in $\mathbb{R}^d$. For the Euclidean norm on $\mathbb{R}^d$, min–max approximants exist to V from $\mathbb{R}^d$ and are essentially unique if V is smooth enough.

It turns out that the proof of Theorem 8.12 takes us into an interesting world of algebraic geometry and approximation along these lines. For example, as we will see the use of an interesting approximation result, the Lojasiewicz inequality in algebraic geometry. The Lojasiewicz inequality, [106], gives an upper bound estimate for the distance between a point in $\mathbb{R}^d$ to the zero set (if non-empty) of a given real analytic function.

Specifically, let $f : U \to \mathbb{R}$ be a real analytic function on an open set U in $\mathbb{R}^d$. Let X' be the zero locus of f. Assume that X' is not empty. Let $X \subset U$ be a compact set. Let x in X. Then there exist constants c, c_1 depending on d with

$$dist(x, X')^{c_1} \leq c|f(x)|. \tag{8.4}$$

An important point regarding Theorem 8.12 (part (1)) is that (8.1) forces the points $\{y_1, ..., y_k\}$ and $\{z_1, ..., z_k\}$ to live on a ellipse. This is useful. The primary reason being that suddenly, we have convexity. A convex set is up to a certain equivalence an ellipse. Min–max approximation and convexity “like each other”. This fact allows Lojasiewicz to be used in a clever way which gives Theorem 8.12 (part (1)) as we show.

Geometrically, the sufficient conditions of Theorem 8.12 (part (2)) provide a geometry on the set E so that again allow Lojasiewicz to be used.

Proof of Theorem 8.12 (part (1). Let us suppose we can find points $y'_1, ..., y'_k$ distinct and $z'_1, ..., z'_k$ distinct both in $\mathbb{R}^d$ satisfying the following.

(1) $|y_i - y'_j| \leq c\delta^{c_1}$, for $1 \leq i, j \leq k$.
(2) $|z_i - z'_j| \leq c_2\delta^{c_3}$, for $1 \leq i, j \leq k$.

(3) $|y_i' - y_j'| = |z_i' - z_j'|$ for $1 \leq i, j \leq k$.
(4) Here, the constants c, c_i are small enough, $i = 1, 2, 3$.
(5) From (3), we can choose a Euclidean motion A so that $A(y_i') = z_i'$ for each $1 \leq i \leq k$. $(1-2)$ then give the result.

So, we need to construct the approximation points $y_1', ..., y_k'$ and $z_1', ..., z_k'$. This follows from (8.1) and (8.2) and Lojasiewicz. Here in particular, (8.1) is used to construct the needed function f.

The proof of Theorem 8.12 (part (2)) follows by the same argument as Theorem 8.12 (part (1)). All that is needed is to observe that under the given conditions on the set E, Lojasiewicz may be applied exactly as in Theorem 8.12 (part (1)).

We refer the reader to the references [58, 94] for an interesting and different perspective on these collections of ideas.

8.2 Min-max Optimization and Convexity

In the last section, we spent some time discussing min–max optimization and convexity. Here we study:

$$\inf_{f_1 \in \mathcal{F}} \max_{x \in X} |f(x) - f_1(x)|.$$

$X \subset \mathbb{R}^n$ is a certain compact set and $f : X \to \mathbb{R}$ is defined globally and is a continuous function. The approximation of f is via a family $\mathcal{F}$ of continuous functions $f_1 : X \to \mathbb{R}$.

Theorem 8.13. *[56]. Given a continuous function f on a certain simplex X in $\mathbb{R}^n$ with the following property. The graph of f is either weak convex or weak concave over X. Then an expression for the min-max uniform affine approximation of f on the simplex X has graph given by $f_1 + Y$ where $2Y$ is the non-zero extremum value of $f - f_1$ on X and where f_1 is the secant hyperplane to f through the vertices of the simplex X. f_1 is also the interpolant to f at the $n + 1$ vertices of the simplex X.*

This generalizes the well-known Chebyshev equioscillation theorem for the case of $n = 1$.

9

Building ε-distortions: Near Reflections

9.1 Theorem 9.14

Our main result in this chapter is going to be a finer result than Theorem 2.1 where we assume more on the geometry of the set E [40].

Here is our result.

Theorem 9.14. *Let ε be small enough depending on d. Let $0 < \tau < 1$, $E \subset \mathbb{R}^d$ be a finite set of distinct points with diam$(E) = 1$ and with $|z - z'| \geq \tau$ for all $z, z' \in E$ distinct. Assume that $V_d(E) \leq \eta^d$ where $0 < \eta < c\tau\varepsilon$ for small enough c. Here we recall V_d is given by Definition 7.10. Then, there exists a $c'\varepsilon$-distorted diffeomorphism $f : \mathbb{R}^d \to \mathbb{R}^d$ with the following properties:*

(1) f coincides with an improper Euclidean motion on $\{x \in \mathbb{R}^d : dist(x, E) \geq 20\}$.
(2) f coincides with an improper Euclidean motion A_z on $B(z, \tau/100)$ for each $z \in E$.
(3) $f(z) = z$ for each $z \in E$.

Notice that (2) tells us that the agreement is pointwise z dependent. This is finer that what we have in Theorem 2.1. Also recall that the conditions of the theorem are that (a) the points of E are not too close together (the constant τ), (b) the points of E are close to a hyperplane (the quantity $V_d(E)$ small), (c) that the diameter of E is not too large).

9.2 Proof of Theorem 9.14

We provide now a proof of Theorem 9.14. We do this in three steps.

Near Extensions and Alignment of Data in $\mathbb{R}^n$: Whitney extensions of near isometries, shortest paths, equidistribution, clustering and non-rigid alignment of data in Euclidean space,
First Edition. Steven B. Damelin.

Step 1: We establish the following:

Lemma 9.15. *Let ε be small enough depending on d. Let $0 < \tau < 1$ and $E \subset \mathbb{R}^d$ be a finite set of distinct points. Assume that diam(E) ≤ 1 and $|z - z'| \geq \tau$ for $z, z' \in E$ distinct. Let $A : \mathbb{R}^d \to \mathbb{R}^d$ be an improper Euclidean motion and assume one has $|A(z) - z| \leq \delta$ for all $z \in E$ where $\delta < c\varepsilon\tau$ for small enough c. Then, there exists a ε-distorted diffeomorphism f such that:*

(1) $f(x) = x$ whenever dist$(x, E) \geq 10$, $x \in \mathbb{R}^d$.
(2) $f(x) = x + [z - A(z)]$, $x \in B(z, \tau/10)$, $z \in E$.

Proof: Let $\theta(y)$ be a smooth cutoff function on $\mathbb{R}^d$ such that $\theta(y) = 1$ for $|y| \leq 1/10$, $\theta(y) = 0$ for $|y| \geq 1/5$ and $|\nabla\theta| \leq C$ on $\mathbb{R}^d$. Let

$$f(x) = \sum_{z\in E}(z - A(z))\theta\left(\frac{x - z}{\tau}\right),\ x \in \mathbb{R}^d.$$

Let $x \in \mathbb{R}^d$. We observe that if dist$(x, E) \geq 10$, then $\frac{|x-z|}{\tau} \geq 1/5$ for each $z \in E$ and so $\theta\left(\frac{x-z}{\tau}\right) = 0$ for each $z \in E$. Thus $f(x) = 0$ if dist$(x, E) \geq 10$. Next if $x \in B(z, \tau/10)$ for each $z \in E$, then for a given $z \in E$, say z_i and for $x \in B(z_i, \tau/10)$, $\frac{|x-z_i|}{\tau} \leq 1/10$ and so $\theta\left(\frac{x-z_i}{\tau}\right) = 1$. However since $|z - z'| \geq \tau$ for $z, z' \in E$ distinct, we also have for $x \in B(z, \tau/10), z \in E,\ z \neq z_i$, $\frac{|x-z|}{\tau} \geq 1/5$ and so $\theta\left(\frac{x-z}{\tau}\right) = 0$ for $z \neq z_i$. Thus $f(x) = z - A(z)$ for $x \in B(z, \tau/10)$, $z \in E$. Finally $|\nabla f| \leq \frac{\eta}{C\tau} < c\varepsilon$ where c is small enough. Then the function $x \to x + f(x)$, $x \in \mathbb{R}^d$ is a Slide and thus a ε-distorted diffeomorphism. Thus Lemma 9.15 holds. □

Step 2:

Using Theorem 7.11, there exists an improper Euclidean motion A such that for $z \in E$, $|A(z) - z| \leq c'\eta$.

Step 3:

Lemma 9.15 applies. Let f_1 be a $c''\varepsilon$-distorted diffeomorphism as in the conclusion of Lemma 9.15. Define $f(x) = f_1(A(x))$, $x \in \mathbb{R}^d$ and we are done □

10

ε-distorted diffeomorphisms, *O(d)* and Functions of Bounded Mean Oscillation (BMO)

Up till now, we have spent some time building ε-distorted diffeomorphisms from elements of $SO(d)$ and Euclidean motions. Here ε is small enough. In this chapter, [42], we are going to study the following problem, namely, given a ε-distorted diffeomorphism, how well can one approximate it by elements of $O(d)$? Here ε is small enough. We will study this problem in volume measure.

In studying this problem, we find interesting connections of this problem to the space of functions $f : \mathbb{R}^d \to \mathbb{R}^d$ of bounded mean oscillation (BMO) and the John–Nirenberg inequality. Here, $\text{vol}_d(\cdot)$ is as usual a d-dimensional volume which in this section we write as $\text{vol}(\cdot)$.

We also need for this chapter the following:

(1) Suppose that $f : \mathbb{R}^d \to \mathbb{R}^d$ is a c-distorted diffeomorphism for some small enough c. Then $|(f'(x))^T f'(x) - I| \leq c'c$.
(2) If M is real and symmetric and if $(1-c)I \leq M \leq (1+c)I$ as matrices for some $0 < c < 1$, then $|M - I| \leq c''c$. This follows from working in an orthonormal basis for which M is diagonal.

10.1 BMO

A function $f : \mathbb{R}^d \to \mathbb{R}^d$ is BMO, if there exists a constant $c_{NN} > 0$ depending on d such that for every ball $B \subset \mathbb{R}^d$, there exists a real number $c'_{RB} := c'(B, f)$ with

$$\frac{1}{\text{vol}(B)} \int_B |f(y) - c'_{RB}| dy \leq c_{NN}.$$

The least such constant c_{NN} is denoted by $|f|_{\text{BMO}}$. This norm turns BMO into a Banach space. BMO was first introduced and studied in the context of elasticity.

Near Extensions and Alignment of Data in $\mathbb{R}^n$: Whitney extensions of near isometries, shortest paths, equidistribution, clustering and non-rigid alignment of data in Euclidean space,
First Edition. Steven B. Damelin.

See [71, 72]. (BMO is often called a John–Nirenberg space and may be understood in a probabilistic framework as is apparent from its definition.) $B = B(z, r)$ is as usual the ball of radius r and center z.

The space BMO appears in the following classical result.

10.2 The John–Nirenberg Inequality

The John–Nirenberg inequality asserts the following: let f be in the space BMO and let $B \subset \mathbb{R}^d$ be a ball. Then there exists a real number $c_{RB} := c(B, f)$ and $c' > 0$ depending on d such that

$$\text{vol}\{x \in B : |f(x) - c_{RB}| > cc'|f|_{BMO}\} \leq \exp(-c)\text{Vol}(B),\ c \geq 1. \tag{10.1}$$

As a consequence of (10.1), we have:

$$\left(\frac{1}{\text{vol}(B)} \int_B |f(x) - c_{RB}|^4\right)^{1/4} \leq c'|f|_{BMO}. \tag{10.2}$$

The definitions of BMO and the notion of the BMO norm can be modified so that the definition of BMO allows f to take its values in the space of $D \times D$ matrices. Then the John–Nirenberg inequality and (10.2) hold again.

10.3 Main Results

For all our results below, ε is small enough and depends on d.

Theorem 10.16. *Let $f : \mathbb{R}^d \to \mathbb{R}^d$ be an ε-distorted diffeomorphism. Let $B \subset \mathbb{R}^d$ be a ball. Then, there exists $M_B \in O(d)$ such that*

$$\text{vol}\{x \in B : |f'(x) - M_B| > cc'\varepsilon\} \leq \exp(-c)vol(B),\ c \geq 1. \tag{10.3}$$

Moreover, Theorem 10.16 is sharp (in the sense of small enough volume) by a Slow Twist.

Theorem 10.17. *Let $f : \mathbb{R}^d \to \mathbb{R}^d$ be an ε-distorted diffeomorphism and let $B \subset \mathbb{R}^d$ be a ball. Then there exists $M_B \in O(d)$ such that*

$$\frac{1}{\text{vol}(B)} \int_B |f'(x) - M_B|\, dx \leq c'\varepsilon^{1/2}. \tag{10.4}$$

Theorem 10.18. *Let $f : \mathbb{R}^d \to \mathbb{R}^d$ be an ε-distorted diffeomorphism and let $B \subset \mathbb{R}^d$ be a ball. Then, there exists $M_B \in O(d)$ such that*

$$\left(\frac{1}{\text{vol}(B)}\int_B |f'(x) - M_B|^4\,dx\right)^{1/4} \leq c'\varepsilon^{1/2}. \tag{10.5}$$

Theorem 10.19. *Let $f : \mathbb{R}^d \to \mathbb{R}^d$ be an ε-distorted diffeomorphism and let $B \in \mathbb{R}^d$ be a ball. Then, there exists $M_B \in O(d)$ such that*

$$\frac{1}{\text{vol}(B)}\int_B |f'(x) - M_B|\,dx \leq c'\varepsilon. \tag{10.6}$$

We mention that Theorem 10.19 is a refinement of Theorem 10.17. We also mention that the the work of the paper [42] has interesting applications to music.

10.4 Proof of Theorem 10.17

Proof: (10.4) is preserved by translations and dilations. Hence we may assume without loss of generality that $B = B(0,1)$. We will prove in Theorem 11.21(part 1) that there exists a Euclidean motion A such that $|f(x) - A(x)| \leq c\varepsilon$, $x \in B(0,1)$. Also, our desired conclusion holds for f iff it holds for the composition $A^{-1}of$ (with possibly a different A). Hence, without loss of generality, we may assume that $A = I$. Thus, we have $|f(x) - x| \leq c\varepsilon$, $x \in B(0,1)$.

We write $f(x_1, ..., x_d) = (y_1, ..., y_d)$ where for each i, $1 \leq i \leq d$, $y_i = f_i(x_1, ..., x_d)$, some smooth family $f_i : \mathbb{R}^d \to \mathbb{R}^d$, $1 \leq i \leq d$.

First claim: for each $i = 1, ..., D$,

$$\int_{B(0,1)} \left|\frac{\partial f_i(x)}{\partial x_i} - 1\right| \leq c'\varepsilon.$$

Let B' denote the ball of radius 1 about the origin in $\mathbb{R}^{d-1}$. For this, for fixed $(x_2, ..., x_d) \in B'$, we know that defining $x^+ = (1, x_2.., x_d)$ and $x^- = (-1, .x_2.., x_d)$ we have:

$$|f_1(x^+) - 1| \leq c'\varepsilon$$

and

$$|f_1(x^-) + 1| \leq c'\varepsilon.$$

Consequently,

$$\int_{-1}^{1} \frac{\partial f_1}{\partial x_1}(x_1, ..., x_d)dx_1 \geq 2 - c'\varepsilon.$$

for $(x_2, ..., x_d) \in B'$. Here c' depends on d.

On the other hand, since,

$$(f'(x))^T (f'(x)) \leq (1+\varepsilon)I,$$

we have the inequality for all $(x_1, ..., x_d)$ in $\mathbb{R}^d$,

$$\left|\frac{\partial f_i}{\partial x_i}(x_1, ..., x_d)\right| \leq 1 + c'\varepsilon.$$

Now if c'' (depending on d) is large enough, from the above, we see now that in $\mathbb{R}^d$

$$\left[1 + c''(\varepsilon) - \frac{\partial f_1}{\partial x_1}(x_1, ..., x_d)\right] \geq 0$$

and for $(x_2, ..., x_d) \in B'$,

$$\int_{-1}^{1} \left|1 + c''(\varepsilon) - \frac{\partial f_1}{\partial x_1}(x_1, ..., x_d)\right| dx_1 \leq 10c''\varepsilon.$$

Hence, for $(x_2, ..., x_D) \in B'$

$$\int_{-1}^{1} \left|1 - \frac{\partial f_1}{\partial x_1}(x_1, ..., x_d)\right| dx_1 \leq c'\varepsilon.$$

Noting that $B(0,1) \subset [-1,1] \times B'$, we see that for each $i = 1, 2, ..., d$,

$$\int_{B(0,1)} \left|\frac{\partial f_i}{\partial x_i} - 1\right| dx \leq c'\varepsilon.$$

This is claim 1.

Since

$$(1-\varepsilon)I \leq (f(x))^T(f'(x)) \leq (1+\varepsilon)I,$$

we have for each i

$$\left|\frac{\partial f_i}{\partial x_i}\right| \leq 1 + c'\varepsilon$$

and

$$\text{trace}\left[(f'(x))^T(f'(x))\right] \leq (1 + c'\varepsilon)d.$$

So,

$$\sum_{i,j=1}^{d}\left(\frac{\partial f_i}{\partial x_j}\right)^2 \leq (1+C\varepsilon)d.$$

Therefore, we have:

$$\begin{aligned}
&\sum_{i\neq j}\left(\frac{\partial f_i}{\partial x_j}\right)^2\\
&\leq (1+c'(\varepsilon))d - \sum_{i=1}^{d}\left(\frac{\partial f_i}{\partial x_j}\right)^2\\
&= c'(\varepsilon) + \sum_{i=1}^{D}\left[1-\left(\frac{\partial f_i}{\partial x_j}\right)^2\right].
\end{aligned}$$

Moreover, for each i, we now have

$$\begin{aligned}
&\left|\left(1-\frac{\partial f_i}{\partial x_j}\right)^2\right|\\
&= \left|\left(1-\frac{\partial f_i}{\partial x_j}\right)\right|\left|\left(1+\frac{\partial f_i}{\partial x_j}\right)\right|\\
&\leq 3\left|\left(1-\frac{\partial f_i}{\partial x_j}\right)\right|.
\end{aligned}$$

And so everywhere on $\mathbb{R}^d$,

$$\sum_{i\neq j}\left(\frac{\partial f_i}{\partial x_j}\right)^2 \leq c'(\varepsilon) + 3\sum_{i}\left|\frac{\partial f_i}{\partial x_j}-1\right|.$$

Now integrating, we find that for $i \neq j$

$$\int_{B(0,1)}\left|\frac{\partial f_i}{\partial x_j}\right|^2 dx \leq c'\varepsilon.$$

Consequently, by the Cauchy–Schwartz inequality, we have for $i \neq j$

$$\int_{B(0,1)}\left|\frac{\partial f_i}{\partial x_j}\right| dx \leq c'\varepsilon^{1/2}.$$

Recalling that $f'(x)$ is just the matrix $\frac{\partial f_i}{\partial x_j}$,

$$\int_{B(0,1)} |f'(x) - I|\, dx \leq c'\varepsilon^{1/2}.$$

Thus, we have proved what we need with $A = I$. The proof of Theorem 10.17 is complete. □

10.5 Proof of Theorem 10.18

Theorem 10.18 follows from Theorem 10.17 and (10.2). □

10.6 Proof of Theorem 10.19

We now prove Theorem 10.19.

Proof: We may assume without loss of generality that

$$B = B(0,1).$$

From Theorem 10.18, we know the following: there exists $M_B \in O(d)$ such that

$$\left(\int_{B(0,10)} |f'(x) - M_B|^4\, dx\right)^{1/4} \leq c'\varepsilon^{1/2}.$$

Our desired conclusion holds for f iff it holds for the composition $M_B^{-1} o f$. Hence without loss of generality, we may assume that $M_B = I$. Thus, we have

$$\left(\int_{B(0,10)} |f'(x) - I|^4\, dx\right)^{1/4} \leq c'\varepsilon^{1/2}.$$

Let now
$F(x) = (F_1(x), F_2(x), ..., F_D(x)) = f(x) - x,\ x \in \mathbb{R}^d.$
Then we have:

$$\left(\int_{B(0,10)} |\nabla(F(x))|^4\, dx\right)^{1/4} \leq c'\varepsilon^{1/2}.$$

We know that

$$(1 - c'(\varepsilon))I \leq (f'(x))^T(f'(x)) \leq (1 + c'\varepsilon)I$$

and so

$$\left|(f'(x))^T(f'(x)) - I\right| \leq c'\varepsilon,\ x \in \mathbb{R}^d.$$

Now, in coordinates, $f'(x)$ is the matrix $\left(\delta_{ij} + \frac{\partial F_i(x)}{\partial x_j}\right)$, hence $f(x)^T f'(x)$ is the matrix whose ij entry is

$$\sum_{l=1}^{d}\left(\delta_{li} + \frac{\partial F_l(x)}{\partial x_i}\right)\left(\delta_{lj} + \frac{\partial F_l(x)}{\partial x_j}\right)$$
$$= \left(\delta_{li} + \frac{\partial F_j(x)}{\partial x_i} + \frac{\partial F_i(x)}{\partial x_j}\right)\left(\sum_{l=1}^{d} \frac{\partial F_l(x)}{\partial x_i} + \frac{\partial F_l(x)}{\partial x_j}\right).$$

Thus

$$\left| \frac{\partial F_i}{\partial x_j} + \frac{\partial F_j}{\partial x_i} + \sum_{l=1}^{D} \frac{\partial F_l}{\partial x_i} \frac{\partial F_l}{\partial x_j} \right| \leq c'\varepsilon$$

on $\mathbb{R}^d$, $i, j = 1, \ldots, D$. Using the Cauchy–Schwartz inequality, we then learn the estimate

$$\left| \sum_{l=1}^{D} \frac{\partial F_l}{\partial x_j} + \frac{\partial F_j}{\partial x_i} \right|_{2(B(0,10))} \leq c'\varepsilon.$$

Continuing we make the following claim: there exists, for each i, j, an antisymmetric matrix $M = (M)_{ij}$, such that

$$\left| \frac{\partial F_i}{\partial x_j} - M \right|_{2(B(0,1))} \leq c'\varepsilon.$$

We know that if true, we have

$$|f' - (I + M)|_{2(B(0,1))} \leq c'\varepsilon.$$

We also know that

$$|M| \leq c'\varepsilon^{1/2}$$

and thus,

$$|\exp(M) - (I + M)| \leq c'\varepsilon.$$

Invoking Cauchy–Schwartz, we have

$$\int_B |f'(x) - \exp(M)(x)| \, dx \leq c'\varepsilon.$$

This implies Theorem 10.19 because M is antisymmetric which means that $\exp(M) \in O(d)$. □

So, to prove Theorem 10.19 we need to establish our claim. This follows by the analysis of a certain overdetermined system which is Theorem 10.20.

10.7 An Overdetermined System

We study the following overdetermined system of partial differential equations.

$$\frac{\partial \Omega_i}{\partial x_j} + \frac{\partial \Omega_j}{\partial x_i} = f_{ij}, i, j = 1, \ldots, d \tag{10.7}$$

on $\mathbb{R}^D$. Here, Ω_i and f_{ij} are C^∞ functions on $\mathbb{R}^d$.

Here is:

Theorem 10.20. *Let $\Omega_1,\ldots,\Omega_d$ and f_{ij}, $i,j = 1,\ldots,D$ be smooth functions on $\mathbb{R}^d$. Assume that (10.7) holds and suppose that*

$$||f_{ij}||_{L^2(B(0,4))} \leq 1. \tag{10.8}$$

Then, there exist real numbers Δ_{ij}, $i,j = 1,\ldots,d$ such that

$$\Delta_{ij} + \Delta_{ji} = 0, \forall i,j \tag{10.9}$$

and

$$\left\| \frac{\partial \Omega_i}{\partial x_j} - \Delta_{ij} \right\|_{L^2(B(0,1))} \leq C. \tag{10.10}$$

Proof: From (10.7) we see at once that

$$\frac{\partial \Omega_i}{\partial x_i} = \frac{1}{2} f_{ii}$$

for each i. Now, by differentiating (10.7) with respect to x_j and then summing on j, we see that

$$\Delta \Omega_i + \frac{1}{2} \frac{\partial}{\partial x_i} \left(\sum_j f_{jj} \right) = \sum_j \frac{\partial f_{ij}}{\partial x_j}$$

for each i. Therefore, we may write

$$\Delta \Omega_i = \sum_j \frac{\partial}{\partial x_j} g_{ij}$$

for smooth functions g_{ij} with

$$||g_{ij}||_{L^2(B(0,4))} \leq C.$$

This holds for each i. Let χ be a C^∞ cutoff function on $\mathbb{R}^D$ equal to 1 on $B(0,2)$ vanishing outside $B(0,4)$ and satisfying $0 \leq \chi \leq 1$ everywhere. Now let

$$\Omega_i^{\mathrm{err}} = \Delta^{-1} \sum_j \frac{\partial}{\partial x_j} (\chi g_{ji})$$

and let

$$\Omega_i^* = \Omega_i - \Omega_i^{err}.$$

Then,

$$\Omega_i = \Omega_i^* + \Omega_i^{err} \tag{10.11}$$

each i.

$$\Omega_i^* \tag{10.12}$$

is harmonic on $B(0,2)$ and

$$\left||\nabla\Omega_i^{err}\right||_{L^2(B(0,2))} \leq c. \tag{10.13}$$

We can now write

$$\frac{\partial\Omega_i^*}{\partial x_j} + \frac{\partial\Omega_j^*}{\partial x_i} = f_{ij}^*, i, j = 1, ..., D \tag{10.14}$$

on $B(0,2)$ and with

$$\left||f_{ij}^*\right||_{L^2(B(0,2)} \leq c_1. \tag{10.15}$$

We see that each f_{ij}^* is a harmonic function on $B(0,2)$. Consequently,

$$sup_{B(0,1)}\left|\nabla f_{ij}^*\right| \leq c_2.. \tag{10.16}$$

We thus have for each i, j, k,

$$\frac{\partial^2\Omega_i^*}{\partial x_j\partial x_k} + \frac{\partial^2\Omega_k^*}{\partial x_i\partial x_j} = \frac{\partial f_{ik}^*}{\partial x_j}; \frac{\partial^2\Omega_i^*}{\partial x_j\partial x_k} + \frac{\partial^2\Omega_j^*}{\partial x_i\partial x_k} = \frac{\partial f_{ij}^*}{\partial x_k} \tag{10.17}$$

$$\frac{\partial^2\Omega_j^*}{\partial x_i\partial x_k} + \frac{\partial^2\Omega_k^*}{\partial x_i\partial x_j} = \frac{\partial f_{jk}^*}{\partial x_i}. \tag{10.18}$$

Now adding the first two equations above and subtracting the last, we obtain:

$$2\frac{\partial^2\Omega_i^*}{\partial x_j\partial x_k} = \frac{\partial f_{ik}^*}{\partial x_j} + \frac{\partial f_{ij}^*}{\partial x_k} - \frac{\partial f_{jk}^*}{\partial x_i} \tag{10.19}$$

on $B(0,1)$. Thus we obtain the estimate

$$\left|\frac{\partial^2\Omega_i^*}{\partial x_j\partial x_k}\right| \leq c_3 \tag{10.20}$$

on $B(0,1)$ for each i, j, k. Now for each i, j, let

$$\Delta_{ij}^* = \frac{\partial\Omega_i^*}{\partial x_j}(0). \tag{10.21}$$

We have then

$$\left|\frac{\partial\Omega_i^*}{\partial x_j} - \Delta_{ij}^*\right| \leq c_4 \tag{10.22}$$

on $B(0,1)$ for each i, j and

$$\left\| \frac{\partial \Omega_i}{\partial x_j} - \Delta^*_{ij} \right\|_{L^2(B(0,1))} \leq c_5, . \tag{10.23}$$

We have the estimate

$$\left| \Delta^*_{ij} + \Delta^*_{ji} \right| \leq c_5$$

for each i, j. Hence, there exist real numbers Δ_{ij}, $(i, j = 1, ..., D)$ such that

$$\Delta_{ij} + \Delta_{ji} = 0 \tag{10.24}$$

and

$$\left| \Delta^*_{ij} - \Delta_{ij} \right| \leq C \tag{10.25}$$

for each i, j. Thus we see that

$$\left\| \frac{\partial \Omega_i}{\partial x_j} - \Delta_{ij} \right\|_{L^2(B(0,1))} \leq c_6 \tag{10.26}$$

for each i and j.
We have proved the theorem. □

10.8 Proof of Theorem 10.16

Proof: The proof of Theorem 10.16 follows from (10.2) and Theorem 10.19. □

11

Results: A Revisit of Theorem 2.2 (part (1))

In this chapter we are going to begin to revisit Theorem 2.2 (part (1)) with a collection of finer results using a new geometry on the set E. As part of our finer results, we will study a quantitative relationship $\delta = \exp\left(-\frac{C_K}{\varepsilon}\right)$ where the constant C_K depends on d and the constant K. ε is small enough and depends on both d and the constant K. Recall that the constant K is chosen when d is chosen and controls Card(E). Recall E is finite so K always exists (but is independent of d). (Moving forward if we write C_K, c_K, C'_K ..., then we mean that the constant depends both on d and K) [40].

11.1 Theorem 11.21

We first formulate and prove a comprehensive result which will allow a construction we develop called an η block. (Recall the constant η controls how close the set E is to a hyperplane.) We need, moving forward the following: a function $f : \mathbb{R}^d \to \mathbb{R}^d$ is proper if $\det(f')(x)$ exists and $\det(f')(x) > 0$ for every $x \in \mathbb{R}^d$.

We will prove:

Theorem 11.21.

(1) *Let ε be small enough depending on d. Let $f : \mathbb{R}^d \to \mathbb{R}^d$ be a ε-distorted diffeomorphism. Let $B := B(z, r)$ be a ball in $\mathbb{R}^d$. Then, there exists a Euclidean motion $A = A_B$ such that:*
   (a) $|f(x) - A(x)| \le c\varepsilon r$, $x \in B$. Here c depends on d.
   (b) Moreover, A is proper iff f is proper.

(2) *Let $\{x_0, ..., x_d\}$ with diam $\{x_0, ..., x_d\} \le 1$ and $V_d(x_0, ..., x_d) \ge \eta^d$ where $0 < \eta < 1$ and let $0 < \varepsilon < c'\eta^d$ for a small enough c'. Let $f : \mathbb{R}^d \to \mathbb{R}^d$ be a ε-distorted diffeomorphism. Finally let A^* be the unique affine function that agrees with*

Near Extensions and Alignment of Data in $\mathbb{R}^n$: Whitney extensions of near isometries, shortest paths, equidistribution, clustering and non-rigid alignment of data in Euclidean space,
First Edition. Steven B. Damelin.

f on $\{x_0, ..., x_d\}$. Then f is proper iff A^* is proper. Here, ε is small enough and depends on d and K. c' depends on d.

Proof: We begin with part (1a). Without loss of generality, we may assume that $B(z, r) = B(0, 1)$ and $f(0) = 0$. Let $e_1, ..., e_d \in \mathbb{R}^d$ be unit vectors. Then, $|f(e_i)| = |f(e_i) - f(0)|$. Hence, for each i,

$$(1 + \varepsilon)^{-1} \leq |f(e_i)| \leq (1 + \varepsilon).$$

Also, for $i \neq j$,

$$(1 + \varepsilon)^{-1}\sqrt{2} \leq |f(e_i) - f(e_j)| \leq (1 + \varepsilon)\sqrt{2}.$$

Hence,

$$f(e_i) \cdot f(e_j) = 1/2\left(|f(e_i)|^2 + |f(e_j)|^2 - |f(e_i) - f(e_i)|^2\right)$$

satisfies

$$|f(e_i) \cdot f(e_j) - \delta_{ij}| \leq c\varepsilon$$

for all i, j where δ_{ij} denotes the Kronecker delta and "$\cdot$" denotes the Euclidean dot product. Applying the Gram–Schmidt process to $f(e_1), \dots, f(e_d)$, we obtain orthonormal vectors $e_1^*, ..., e_d^* \in \mathbb{R}^d$ such that $|f(e_i) - e_i^*| \leq c\varepsilon$ for each i. Let A be the (proper or improper) rotation such that $Ae_i = e_i^*$ for each i. Then $f^{**} := A^{-1} o f$ is an ε-distorted diffeomorphism, $f^{**}(0) = 0$ and $|f^{**}(e_i) - e_i| \leq c\varepsilon$ for each i. Now let $x = (x_1, ..., x_d) \in B(0, 1)$ and let $y = (y_1, ..., y_d) = f^{**}(x)$. Then $2x_i = 1 + |x - 0|^2 - |x - e_i|^2$ and also $2y_i = 1 + |y - 0|^2 - |y - e_i|^2$ for each i. Hence, by the above-noted properties of f^{**}, we have $|y_i - x_i| \leq c\varepsilon$. Then, $|f^{**}(x) - x| \leq c\varepsilon$ for all $x \in B(0, 1)$, i.e., $|f(x) - A(x)| \leq c\varepsilon$ for all $x \in B(0, 1)$. Thus, we have proved (a) but not yet (b). For each (z, r), (a) provides a Euclidean motion $A_{(z,r)}$ such that $|f(x) - A_{(z,r)}(x)| \leq c\varepsilon r$ for $x \in B(z, r)$. Now for r small enough, we have using the mean value theorem for vector valued functions and the substitution rule with Jacobian determinants as expansions of volumes,

$$|f(x) - [f(z) + f'(z)(x - z)]| \leq c\varepsilon r, \ x \in B(z, r).$$

Hence,

$$|A_{(z,r)}(x) - [f(z) + f'(z)(x - z)]| \leq c\varepsilon r, \ x \in B(z, r).$$

Thus we have established for small enough r that $A_{(z,r)}$ is proper iff $\det f'(z) > 0$ i.e., iff f is proper. Observe that we have $|f(x) - A_{(z,r/2)}(x)| \leq c\varepsilon r$ for $x \in B(z, r/2)$. Thus $|A_{(z,r)} - A_{(z,r/2)}(x)| \leq c\varepsilon r$ for $x \in B(z, r/2)$. Hence $A_{(z,r)}$ is proper iff $A_{(z,r/2)}$ is proper. Thus we may deduce that for all r, $A_{(z,r)}$ is proper iff f is proper. This completes the proof of (b) and part (1) of Theorem 11.21. We now prove part (2).

Without loss of generality, we may assume that $x = 0$ and $f(x_0) = 0$. Then A^* is linear, not just affine. By part (1a), there exists a Euclidean motion $A_{(0,1)}$ such that

$$|f(x) - A_{(0,1)}(x))| \leq c\varepsilon$$

for all $x \in B(0, 1)$ and f is proper iff $A_{(0,1)}$ is proper. We know that

$$|A^*(x_i) - A_{(0,1)}(x_i)| \leq c\varepsilon, i = 0, 1, ...d.$$

since $A^*(x_i) = f(x_i)$ and also since $x_i \in B(0, 1) = B(x_0, 1)$. (The latter follows because diam $\{x_0, ..., x_d\} \leq 1$.) In particular, $|A_{(0,1)}(0)| \leq c\varepsilon$ since $x_0 = 0$. Hence,

$$|A^*(x_i) - [A_{(0,1)}(x_i) - A_{(0,1)}(0)]| \leq c'\varepsilon$$

for $i = 1, ..., d$. Now, the function $x \mapsto Ax := A_{(0,1)}(x) - A_{(0,1)}(0)$ is a proper or improper rotation and $\det(A) > 0$ iff $A_{(0,1)}$ is proper iff f is proper. Thus, we have the following:

- $|(A^* - A)x_i| \leq c'\varepsilon, i = 1, ..., d$.
- $|x_1 \wedge ... \wedge x_D| \geq c\eta^d$ (see Section (7.6)).
- $\det(A) > 0$ iff f is proper.
- A is a proper or improper rotation.

Now let L be the linear function that sends the ith unit vector e_i to x_i. Then the entries of L are at most 1 in absolute value since each x_i belongs to $B(0, 1)$. Letting $|\cdot|$ be understood, appropriately, we have from the discussion in Section (7.6) and the above,

$$|\det(L)| = |x_1 \wedge ... \wedge x_d| \geq c\eta^d.$$

Hence by Cramer's rule, $|L^{-1}| \leq c\eta^{-d}$. We now have for each i,

$$|(A^* - A)Le_i| = |(A^* - A)x_i| \leq c'\varepsilon.$$

Hence,

$$|(A^* - A)L| \leq c''\varepsilon$$

and thus

$$|A^* - A| \leq c|(A^* - A)L||L^{-1}| \leq c\varepsilon\eta^{-d}.$$

Since A is a (proper or improper) rotation, it follows that

$$|A^*A^{-1} - I| \leq c\varepsilon\eta^{-d}.$$

Therefore, if $\varepsilon\eta^{-d} \leq c'$ for small enough c', then A^*A^{-1} lies in a small enough neighborhood of I and, therefore, $\det(A^*A^{-1}) > 0$. Hence $\det(A^*)$ and $\det(A)$ have the same sign. Thus, $\det(A^*) > 0$ iff f is proper. So, we have proved (2) and the Theorem. □

We are now ready for:

11.2 η blocks

Definition 11.22. *Let $E \subset \mathbb{R}^d$ be a finite collection of distinct points. Let $f : E \to \mathbb{R}^d$ and let $0 < \eta < 1$. A positive (resp. negative) η-block for f is a $d+1$ tuple $(x_0, ..., x_d) \in \mathbb{R}^d$ such that the following two conditions hold:*

(1) $V_d(x_0, ..., x_d) \geq (\leq)\eta^d diam(x_0, ..., x_d)$.
(2) Let A^* be the unique affine function which agrees with f on E. See Theorem 11.21 ((part (2)). Then we assume that $A*$ is proper(improper). (Note that if the function $A*$ is not invertible then $(x_0, ..., x_d)$ is not an η block.)

We now have the following collection:

Theorem 11.23. *Fix K. Let $\phi : E \to \mathbb{R}^d$ with $E \subset \mathbb{R}^d$ a finite set of distinct points and let $0 < \tau < 1$.*

- *Assumptions on the set E: $diam(E) \leq 1$, $|x-y| \geq \tau$, for any $x, y \in E$ distinct, $card(E) \leq K$.*
- *Assumptions on parameters: $\delta \leq c_K \tau^{C_K}$, C_K is large enough and c_K is small enough. δ is small enough and depends on d and K.*
- *Assumption on ϕ: ϕ satisfies (2.1). Recall*

$$|x-y||(1+\delta)^{-1} \leq |\phi(x) - \phi(y)| \leq (1+\delta)|x-y|, \ x, y \in E.$$

Then there exists a $C'_K \delta^{1/C_K} \tau^{-1}$-distorted diffeomorphism $\Phi : \mathbb{R}^d \to \mathbb{R}^d$ with the following properties:

- *$\Phi = \phi$ on E.*
- *Φ agrees with a Euclidean motion on $\{x \in \mathbb{R}^d : \mathrm{dist}(x, E) \geq 100\}$.*
- *For each $z \in E$, Φ agrees with a Euclidean motion A_z on $B(z, \tau/100)$.*

We now worry about whether the function Φ in Theorem 11.23 is proper or improper. Thus, we have:

Theorem 11.24. *Fix K. Let $\phi : E \to \mathbb{R}^d$ with $E \subset \mathbb{R}^d$ a finite collection of distinct points and let $0 < \tau, \eta < 1$. Let $\varepsilon > 0$ depend on d and K and be small enough.*

- *Assumptions on the set E: $diam(E) = 1$, $|x-y| \geq \tau$, for any $x, y \in E$ distinct, $card(E) \leq K$, $V_d(E) \leq \eta^d$. (Recall here that $V_d(E)$ is small; so this means E is close to a hyperplane.)*
- *Assumption on ϕ: ϕ satisfies (2.1).*
- *Assumptions on η, τ: $0 < \eta < c\varepsilon\tau$, c small enough, $C'_K \delta^{\frac{1}{C_K}} \leq \varepsilon\tau$, C_K and C'_K large enough.*

Then, there exists a proper $c'\varepsilon$-distorted diffeomorphism (c' depending on d) $\Phi : \mathbb{R}^d \to \mathbb{R}^d$ with the following properties:

- $\Phi = \phi$ *on* E.
- Φ *agrees with a proper Euclidean motion on* $\{x \in \mathbb{R}^d : \text{dist}(x, E) \geq 1000\}$.
- *For each* $z \in E$, Φ *agrees with a proper Euclidean motion* A_z *on* $B(z, \tau/100)$.

Theorem 11.25. *Fix K. Let* $\phi : E \to \mathbb{R}^D$ *with* $E \subset \mathbb{R}^d$ *a finite collection of distinct points,* $0 < \tau, \eta < 1$. *Let* $\varepsilon > 0$ *depend on d and the constant K and be small enough.*

- *Assumptions on the set E:* $\text{diam}(E) = 1$, $|x - y| \geq \tau$, *for any* $x, y \in E$ *distinct, card*$(E) \leq K$, $V_d(E) \geq \eta^D$.
- *Assumption on* ϕ: ϕ *satisfies (2.1) with no negative* η *blocks.*
- *Assumptions on parameters* $\varepsilon, \tau, \eta, \delta$.
- *(1):* $C_K' \delta^{1/C_K} \tau^{-1} < \eta^d < 1$ *with* C_K *and* C_K' *large enough.*
- *(2):* $C_K' \delta^{1/C_K} \tau^{-1} < \varepsilon$. *Here the constants* C_K *and* C_K' *are the same as (1).*

Then, there exists a proper $c\varepsilon$*-distorted diffeomorphism (c depends on d)* $\Phi : \mathbb{R}^d \to \mathbb{R}^d$ *with the following properties:*

- $\Phi = \phi$ *on* E.
- Φ *agrees with a proper Euclidean motion* A_∞ *on* $\{x \in \mathbb{R}^d : \textit{dist}(x, E) \geq 1000\}$.
- *For each* $z \in E$, Φ *agrees with a proper Euclidean motion* A_z *on* $B(z, \tau/1000)$.

From Theorem 11.24 and Theorem 11.25 we obtain immediately:

Theorem 11.26. *Fix K. Let* $\phi : E \to \mathbb{R}^d$ *with* $E \subset \mathbb{R}^d$ *finite and let* $0 < \tau, \eta < 1$. *Let* $\varepsilon > 0$ *depend on d and K and be small enough. We make the following assumptions:*

- *Assumptions on the set E:* $\text{diam}(E) = 1$, $|x - y| \geq \tau$, *for any* $x, y \in E$ *distinct, card*$(E) \leq K$.
- *Assumption on* ϕ: ϕ *satisfies (2.1) with no negative* η *blocks.*
- *Assumptions on parameters:* $0 < \eta < c\varepsilon\tau$, *c small enough depending on d,* $c_K \delta^{1/c_K'} \tau^{-1} \leq \min(\varepsilon, \eta^d)$, c_K *and* c_K' *large enough.*

Then, there exists a proper $c\varepsilon$*-distorted diffeomorphism* $\Phi : \mathbb{R}^d \to \mathbb{R}^d$ *with the following properties:*

- $\Phi = \phi$ *on* E.
- Φ *agrees with a proper Euclidean motion* A_∞ *on* $\{x \in \mathbb{R}^d : \text{dist}(x, E) \geq 1000\}$.
- *For each* $z \in E$, Φ *agrees with a proper Euclidean motion* A_z *on* $B(z, \tau/1000)$.

We remark that it follows immediately from Theorem 11.26 that $A_z = \Phi = \phi$ for each $z \in E$; (indeed, if $z \in E$, then trivially $z \in B(z, \tau/1000)$ and also $\Phi = \phi$ for each $z \in E$ and so $\Phi = \phi = A_z$ on E).

One main result in this chapter is then the following Theorem 11.27.

Theorem 11.27. *Fix K. There exist positive constants c_K, c'_K, c''_K such that the following holds: set $\eta = \exp(-c'_K/\varepsilon)$ and $\delta = \exp(-c''_K/\varepsilon)$ with $0 < \varepsilon < c_K$. Let $S \subset \mathbb{R}^d$ be a finite collection of distinct points with* $\text{card}(S) \leq K$. *Let $\phi : S \to \mathbb{R}^D$ satisfy (2.1). Then if ϕ has no negative η block, there exists a proper ε-distorted diffeomorphism $\Phi : \mathbb{R}^d \to \mathbb{R}^d$ such that $\phi = \Phi$ on S and Φ agrees with a proper Euclidean motion on*

$$\left\{x \in \mathbb{R}^d : \text{dist}(x, S) \geq 10^4 \text{diam}(S)\right\}.$$

Theorem 11.28. *Fix K. There exist positive constants c_K, c'_K, c''_K such that the following holds: set $\eta = \exp(-c'_K/\varepsilon)$ and $\delta = \exp(-c''_K/\varepsilon)$ with $0 < \varepsilon < c_K$. Let $S \subset \mathbb{R}^d$ be a finite collection of distinct points with* $\text{card}(S) \leq K$. *Let $\phi : S \to \mathbb{R}^d$. Then if ϕ has a negative η block, ϕ cannot be extended to a proper δ-distorted diffeomorphism on $\mathbb{R}^d$.*

Theorem 11.29. *Fix K. Let $S \subset \mathbb{R}^d$ be a finite set of distinct points with* $\text{card}(S) \leq$ $d+1$. *There exist c, c′ depending on d such that the following holds. Set $\delta = \exp\left(\frac{-c'}{\varepsilon}\right)$ with $0 < \varepsilon < c$ and let $\phi : S \to \mathbb{R}^d$ satisfy (2.1). Then there exists a ε-distorted diffeomorphism $\Phi : \mathbb{R}^d \to \mathbb{R}^d$ such that $\Phi = \phi$ on S. ε is small enough depending on d.*

Theorem 11.30. *Fix K. There exist positive constants c_K, c'_K, c''_K such that the following holds: set $\eta = \exp(-c'_K/\varepsilon)$ and $\delta = \exp(-c''_K/\varepsilon)$ with $0 < \varepsilon < c_K$. Let $S \subset \mathbb{R}^d$ be a finite collection of distinct points with* $\text{card}(S) \leq K$. *Let $\phi : S \to \mathbb{R}^d$. Suppose ϕ has a positive η block say $\{x_0, ..., x_d\}$ and a negative η block, $\{y_0, ..., y_d\}$. Then the restriction of ϕ to $\{x_0, ..., x_d, y_0, ..., y_d\}$ cannot extend to a δ-distorted diffeomorphism $\Phi : \mathbb{R}^d \to \mathbb{R}^d$.*

11.3 Finiteness Principle

As a consequence of Theorem 11.27, we now have Theorem 11.31 which is a finiteness principle.

Theorem 11.31. *Fix K. Let $S \subset \mathbb{R}^d$ be a finite set of distinct points. There exist positive constants c_K, c'_K such that the following holds. Set $\delta = \exp(-\frac{c'_K}{\varepsilon})$ with $0 <$ $\varepsilon < c_K$. Let $\phi : S \to \mathbb{R}^d$. Suppose that for any finite set $S_0 \subset E$ with at most $2d + 2$ points, there exists a δ-distorted diffeomorphism $\Phi^{S_0} : \mathbb{R}^d \to \mathbb{R}^d$ such that $\Phi^{S_0} = \phi$ on S_0. Then, there exists an ε-distorted diffeomorphism $\Phi : \mathbb{R}^d \to \mathbb{R}^d$ such that $\Phi = \phi$ on S.*

12

Proofs: Gluing and Whitney Machinery

12.1 Theorem 11.23

We begin with Theorem 11.23.

Proof: Using the Lojasiewicz inequality, (see Chapter 8), there exists a Euclidean motion A for which we have

$$|\phi(x) - A(x)| \leq c_K \delta^{1/C_K},\ x \in E.$$

Without loss of generality, we may replace ϕ by $\phi o A^{-1}$. Hence, we may suppose that

$$|\phi(x) - x| \leq C'_K \delta^{1/C_k},\ x \in E.$$

Now we will employ a similar technique to the proof of Lemma 9.15.

Let $\theta(y)$ be a smooth cut off function on $\mathbb{R}^d$ such that $\theta(y) = 1$ for $|y| \leq 1/100$, $\theta(y) = 0$ for $|y| \geq 1/50$ and with $|\nabla\theta(y)| \leq C$ for all y. Then set

$$f(x) = \sum_{z \in E} (\phi(z) - z)\theta(x - z/\tau),\ x \in \mathbb{R}^D.$$

The summands are smooth and have pairwise disjoint supports and thus f is smooth. As in the proof of Lemma 9.15, $f(x) = 0$ for $\text{dist}(x, E) \geq 100$, $f(x) = \phi(z) - z$ for $x \in B(z, \frac{\tau}{100})$, $z \in E$ and $|\nabla f(x)| \leq C'_K \delta^{1/C_K} C\tau^{-1}$. If $CC'_K \delta^{1/C_K}\tau^{-1}$ is small enough the function $\Phi(x) = f(x) + x$ is a Slide and thus Φ is a $C''_K \delta^{1/C_K}\tau^{-1}$ distorted diffeomorphism and has all the desired properties. Thus, we are done. □

We now worry about whether the function Φ in Theorem 11.23 is proper or improper. Thus we have Theorem 11.24.

Near Extensions and Alignment of Data in $\mathbb{R}^n$: Whitney extensions of near isometries, shortest paths, equidistribution, clustering and non-rigid alignment of data in Euclidean space,
First Edition. Steven B. Damelin.

Proof: Start with Φ from Theorem 11.23. If Φ is proper, then we are done. (Note that $C'_K\delta^{1/C_K}\tau^{-1} < \varepsilon$.) If Φ is improper, then Theorem 9.14 applies; letting Ψ be as in Theorem 9.14, we see that $\Phi o \Psi$ satisfies all the assertions of Theorem 11.24. □

Next we give the proof of Theorem 11.25:

Proof: We apply Theorem 11.23. The function Φ in Theorem 11.23 is a $C_K\delta^{1/\rho_K}\tau^{-1}$ distorted diffeomorphism; hence is a $C\varepsilon$-distorted diffeomorphism. If Φ is proper, then it satisfies all the conditions needed and we are done. Thus let us check Φ is proper. By hypothesis, we can find $z_1, ..., z_d \in E$ such that

$$V_d(z_0, ..., z_d) \geq \eta^d.$$

Let T be the one and only affine function that agrees with ϕ on $\{z_0, ..., z_d\}$. Since ϕ has no negative η blocks (by hypothesis), we know that T is proper. Applying Theorem 11.21 (b) with δ replaced by $C_K\delta^{1/\rho_K}\tau^{-1}$, we find that Φ is proper as needed. Note that Theorem 11.21 applies here since we assumed that $C_K\delta^{1/\rho_K}\tau^{-1} < \eta^d$ for large enough C_K and ρ_K depending only on K and d. The proof of Theorem 11.25 is complete. □

Combining Theorem 11.24 and Theorem 11.25 we are able to give the proof of Theorem 11.26.

Proof: If $V_d(E) \leq \eta^d$, then Theorem 11.26 follows from Theorem 11.24. If instead, $V_d(E) > \eta^d$, then Theorem 11.26 follows from Theorem 11.25. □

12.2 The Gluing Theorem

Given a finite set E of distinct points with some geometry and a δ distortion ϕ on E, we have investigated in detail up to now how to produce smooth ε-distortions which agree with ϕ on the set E and which agree with Euclidean motions. We need now to "glue" these results together. This will be the subject of this section.

We prove:

Theorem 12.32. *Let E be finite set of distinct points, $\varepsilon > 0$, $0 < \tau < 1$, $\phi : E \to \mathbb{R}^d$ and suppose $|x-y| \geq \tau > 0$ for $x, y \in E$ distinct. Suppose also that*

$$1/2|x-y| \leq |\phi(x) - \phi(y)| \leq 2|x-y|$$

for $x, y \in E$ distinct. For $i = 1, ..., 4$ and $z \in E$, define

$$B_i(z) = B\left(z, \exp\left((i-5)/\varepsilon\right)\tau\right).$$

For each $z \in E$, suppose we are given a $C\varepsilon$-distorted diffeomorphism Φ_z such that $\Phi_z(z) = \phi(z)$ on E and Φ_z agrees with a proper Euclidean motion A_z outside $B_1(z)$ for each $z \in E$. Moreover, suppose we are given a $C\varepsilon$-distorted diffeomorphism Ψ

such that $\phi = \Psi$ on E and Ψ agrees with a proper Euclidean motion A_z^* in $B_4(z)$ for each $z \in E$. Then there exists a $C\varepsilon$-distorted diffeomorphism Φ such that:

- *$\Phi = \Phi_z$ in $B_2(z)$ for $z \in E$ (in particular $\Phi = \phi$ on E) and*
- *$\Phi = \Psi$ outside $\cup_{z\in E} B_3(z)$.*

Proof: We first investigate how well $A_z(z)$ approximates $A_z^*(z)$. Let $z \in E$. Then $A_z^*(z) = \Psi(z) = \phi(z)$ since $z \in B_4(z)$. Moreover, for any $x \in \mathbb{R}^d$ such that $|x-z| = \exp(-4/z)\tau$, we have $x \notin B_1(z)$, hence $\Phi_z(x) = A_z(x)$. We recall that Φ_z is a $C\varepsilon$ distorted-diffeomorphism and that $\Phi_z(z) = \phi(z)$. Thus,

$$(1 + C\varepsilon)^{-1}|x - z| \leq |\Phi_z(x) - \Phi_z(z)| \leq (1 + C\varepsilon)|x - z|$$

i.e.,

$$(1 + C\varepsilon)^{-1} \exp(-4/\varepsilon)\tau \leq |A_z(x) - \phi(z)| \leq (1 + C\varepsilon) \exp(-4/\varepsilon)\tau.$$

This holds whenever $|x - z| = \exp(-4/\varepsilon)\tau$. Since A_z is a Euclidean motion, it follows that

$$|A_z(z) - \phi(z)| \leq C\varepsilon \exp(-4/\varepsilon)\tau.$$

Recalling that $A_z^* = \phi(z)$, we conclude that for $z \in E$,

$$|A_z(z) - A_z^*(z)| \leq C\varepsilon \exp(-4/\varepsilon)\tau.$$

Also, both A_z and A_z^* are proper Euclidean motions and so we obtain for each $z \in E$, a $C\varepsilon$-distorted diffeomorphism Φ_z^* such that:

- Φ_z^* agrees with A_z on $B_2(z)$.
- Φ_z^* agrees with A_z^* outside $B_3(z)$.

Let us define a function $\Phi : \mathbb{R}^d \to \mathbb{R}^d$ in overlapping regions as follows:

- $\Phi = \Phi_z$ in $B_2(z)$ for $z \in E$.
- $\Phi = \Phi_z^*$ in $B_4(z) \setminus B_1(z)$, $z \in E$.
- $\Phi = \Psi$ in $\mathbb{R}^d \setminus \cup_{z\in E} B_3(z)$.

Let us check that the above definitions of Φ in overlapping regions are mutually consistent.

- On $B_2(z) \cap [B_4(z') \setminus B_1(z')]$, $z, z' \in E$: To have a non-empty intersection, we must have $z' = z$ (since otherwise $|z - z'| \geq \tau$). In the region in question, $\Phi_z^* = A_z$ (since we are in $B_2(z)$)$=\Phi_z$ (since we are outside $B_1(z)$).
- On $[B_4(z) \setminus B_1(z)] \cap [\mathbb{R}^d \setminus \cup_{z'\in E} B_3(z')]$, $z \in E$. $\Psi = A_z^*$ (since we are in $B_4(z)$) $= \Phi_z^*$ (since we are outside $B_3(z)$).
- Note that the balls $B_2(z)$, $z \in E$ are pairwise disjoint as are the regions $B_4(z) \setminus B_1(z)$, $z \in E$ since $|z - z'| \geq \tau$ for $z, z' \in E$ distinct.

Moreover, $B_2(z) \cap [\mathbb{R}^d \setminus \cup_{z' \in E} B_3(z')] = \emptyset$. Thus, we have already discussed all the non-empty intersections of the various regions in which Φ was defined. This completes the verification that Φ is defined consistently.

Since Ψ, Φ_z, Φ_z^* (each $z \in E$) are $C\varepsilon$-distorted diffeomorphisms, we conclude that $\Phi : \mathbb{R}^d \to \mathbb{R}^d$ is a smooth function and that

$$(1 + C'\varepsilon)^{-1} \le (\Phi'(x)^T(\Phi'(x)) \le 1 + C'\varepsilon,\ x \in \mathbb{R}^d.$$

We have also $\Phi = \Phi_z$ on $B_2(z)$ for each $z \in E$ and $\Phi = \Psi$ outside $\cup_{z \in E} B_3(z)$ by definition of Φ. To complete the proof of the gluing theorem, it remains only to check that $\Phi : \mathbb{R}^d \to \mathbb{R}^d$ is one-to-one and onto. To see this, we argue as follows. Recall that the A_z and A_z^* are Euclidean motions and that

$$|A_z - A_z^*| \le C\varepsilon \exp(-4/\varepsilon)\tau = C\varepsilon \text{radius}(B_1(z)),\ z \in E.$$

Outside $B_2(z)$, we have $\Phi_z = A_z$. Since $\Phi_z : \mathbb{R}^d \to \mathbb{R}^d$ is one-to-one and onto, it follows that $\Phi_z : B_2(z) \to A_z(B_2(z))$ is one-to-one and onto. Consequently, since $\Phi = \Phi_z$ on $B_2(z)$, we have that:

- $\Phi : B_2(z) \to A_z(B_2(z))$ is one-to-one and onto for each $z \in E$.

Next, recall that $\Phi_z^* = A_z$ on $B_2(z)$, in particular

$$\Phi_z^* : B_2(z) \to A_z(B_2(z))$$

is one-to-one and onto. Also, $\Phi_z^* : \mathbb{R}^d \to \mathbb{R}^d$ is one-to-one and onto and $\Phi_z^* = A_z^*$ outside $B_4(z)$ so it follows that

$$\Phi_z^* : B_4(z) \to A_z^*(B_4(z))$$

is one-to-one and onto. Consequently

$$\Phi_z^* : B_4(z) \setminus B_2(z) \to A_z^*(B_4(z)) \setminus A_z(B_2(z))$$

is one-to-one and onto. Since $\Phi = \Phi^*$ on $B_4(z) \setminus B_2(z)$, we conclude that

- $\Phi : B_4(z) \setminus B_2(z) \to A_z^*(B_4(z)) \setminus A_z(B_2(z))$ is one-to-one and onto for $z \in E$.

Next, recall that $\Psi : \mathbb{R}^d \to \mathbb{R}^d$ is one-to-one and onto and that $\Psi = A_z^*$ on $B_4(z)$ for each $z \in E$. Hence,

$$\Psi : \mathbb{R}^d \setminus \cup_{z \in E} B_4(z) \to \mathbb{R}^d \setminus \cup_{z \in E} A_z^*(B_4(z))$$

is one-to-one and onto. Since $\Phi = \Psi$ on $\mathbb{R}^D \setminus \cup_{z \in E} B_4(z)$, we conclude that

- $\Phi : \mathbb{R}^d \setminus \cup_{z \in E} B_4(z) \to \mathbb{R}^d \setminus \cup_{z \in E} A_z^*(B_4(z))$

  is one-to-one and onto.

Recall that $B_2(z) \subset B_4(z)$ for each $z \in E$ and that the balls $B_4(z), z \in E$ are pairwise disjoint. Hence, the following sets constitute a partition of $\mathbb{R}^d$:

- $B_2(z)$ (all $z \in E$); $B_4(z) \setminus B_2(z)$ (all $z \in E$); $\mathbb{R}^d \setminus \cup_{z\in E} B_4(z)$.

Moreover, we recall that A_z, A_z^* are Euclidean motions, $B_2(z), B_4(z)$ are balls centered at z with radii $\exp(-3/\varepsilon)\tau$ and $\exp(-1/\varepsilon)\tau$ respectively and

$$|A_z(z) - A_z^*(z)| \leq C\varepsilon \exp(-4/\varepsilon)\tau.$$

It follows that $A_z(B_2(z)) \subset A_z^*(B_4(z))$ for $z \in E$. Moreover, $A_z^* = \phi(z)$ for $z \in E$. For $z, z' \in E$ distinct, we have

$$|\phi(z) - \phi(z')| \geq 1/2|z - z'| \geq 1/2\tau.$$

Since, $A_z^*(B_4(z))$ is a ball of radius $\exp(-1/\varepsilon)\tau$ centered at $\phi(z)$ for each $z \in E$, it follows that the balls $A_z^*(B_4(z))\,(z \in E)$ are pairwise disjoint. Therefore, the following sets constitute a partition of $\mathbb{R}^d$:

- $A_z(B_2(z))(z \in E)$, $A_z * (B_4(z)) \setminus A_z(B_2(z))(z \in E)$, $\mathbb{R}^d \setminus \cup_{z\in E} A_z^*(B_4(z))$.

We conclude that $\Phi : \mathbb{R}^d \to \mathbb{R}^d$ is one-to one and onto. The proof of the gluing theorem is complete. □

12.3 Hierarchical Clusterings of Finite Subsets of $\mathbb{R}^d$ Revisited

We are almost ready for the proofs of Theorem 11.27 and Theorem 11.28. We need one more piece of machinery, hierarchical clusterings which we have already seen in Chapter 5. We need the following modified form of the result found there, the proof of which is identical.

Lemma 12.33. *Fix K (independent of d). Let $E \subset \mathbb{R}^d$ with $2 \leq card(E) \leq K$. Let $\varepsilon > 0$. Then there exists τ satisfying*

$$\exp(-C_K/\varepsilon)\text{diam}(E) \leq \tau \leq \exp(-1/\varepsilon)\text{diam}(E)$$

and a partition of E into subsets $E_\nu(\nu = 1, ..., \nu_{(max)})$ with the following properties:

- $\text{card}(E_\nu) \leq K - 1$, $\forall \nu$.
- $\text{diam}(E_\nu) \leq \exp(-5/\varepsilon)\tau$, $\forall \nu$.
- $\text{dist}(E_\nu, E_{\nu'}) \geq \tau$, $\forall \nu$.

12.4 Proofs of Theorem 11.27 and Theorem 11.28

We begin with:

Proof of Theorem 11.27 We use induction on K. If $K = 1$, the theorem holds trivially. For the induction step we will fix $K \geq 2$ and assume that our result holds for $K-1$. We now establish the theorem for the given K. Thus, we are making the following inductive assumptions. For suitable constants $c_{\text{old}}, C'_{\text{old}}, C''_{\text{old}}$ depending only on d, K the following holds. Inductive hypothesis: suppose that $0 < \varepsilon < c_{\text{old}}$, define $\eta_{\text{old}} = \exp(-C'_{\text{old}}/\varepsilon)$ and $\delta_{\text{old}} = \exp(-C''_{\text{old}}/\varepsilon)$. Let $\phi^* : S^* \to \mathbb{R}^d$ with $S^* \subset \mathbb{R}^d$ and $\text{card}(S^*) \leq K-1$. Suppose

$$(1+\delta_{\text{old}})^{-1}|x-y| \leq |\phi^*(x) - \phi^*(y)| \leq (1+\delta_{\text{old}})|x-y|,\ x, y \in S.$$

Then the following holds: if ϕ^* has no negative η_{old} block, then there exists a proper ε-distorted diffeomorphism $\Phi^* : \mathbb{R}^d \to \mathbb{R}^d$ such that $\phi^* = \Phi^*$ on S and Φ^* agrees with a proper Euclidean motion on

$$\{x \in \mathbb{R}^d : \text{dist}(x, S^*) \geq 10^4 \text{diam}(S^*)\}.$$

Now let L, L', L'' be positive constants to be fixed later. (Eventually we will let them depend on d and K but not yet.) Now suppose that

(1) $0 < \varepsilon < L$
(2) we set $\eta = \exp(-L'/\varepsilon)$
(3) and we set $\delta = \exp(-L''/\varepsilon)$
(4) let $\phi : S \to \mathbb{R}^d$
(5) where $S \subset \mathbb{R}^d$
(6) $\text{card}(S) = K$ and
(7) $(1+\delta)^{-1}|x-y| \leq |\phi(x) - \phi(y)| \leq (1+\delta)|x-y|,\ x, y \in S$
(8) suppose that ϕ has a negative η block.

We will construct a proper $C\varepsilon$-distorted diffeomorphism Φ that agrees with ϕ on S and with a proper Euclidean motion away from S. To do so, we first apply the clustering lemma, Lemma 12.33. Recall that $\text{card}(S) = K \geq 2$ so the clustering lemma applies. Let τ and $S_\nu (\nu = 1, \ldots, \nu_{\max})$ be as in the clustering lemma. Thus,

(9) S is the disjoint union of $S_\nu (\nu = 1, \ldots, \nu_{\max})$.
(10) $\text{card}(S_\nu) \leq K-1$ for each $\nu (\nu = 1, \ldots, \nu_{\max})$.
(11) $\text{diam} S_\nu \leq \exp(-5/\varepsilon)\tau$ for each $\nu (\nu = 1, \ldots, \nu_{\max})$.
(12) $\text{dist}(S_\nu, S_{\nu'}) \geq \tau \text{diam}(S)$, for $\nu \neq \nu'$, for each $\nu, \nu' (\nu, \nu' = 1, \ldots, \nu_{\max})$.
(13) $\exp(-C_K/\varepsilon)\text{diam}(S) \leq \tau \leq \exp(-1/\varepsilon)\text{diam}(S)$.
(14) Assuming that $L' > C'_{\text{old}}$ and $L'' > C''_{\text{old}}$, we see that $\eta < \eta_{\text{old}}$ and $\delta < \delta_{\text{old}}$. Hence by (7) and (8) we have:

(15) $\phi|S_\nu$ does not have an η_{old} block and

$$(1+\delta_{\text{old}})^{-1}|x-y| \leq |\phi(x)-\phi(y)| \leq (1+\delta_{\text{old}})|x-y|,\ x,y \in S_\nu$$

Consequently (10) and the induction hypothesis
(16) produce a proper ε-distorted diffeomorphism $\Phi_\nu : \mathbb{R}^d \to \mathbb{R}^d$ such that
(17) $\Phi_\nu = \phi$ on S_ν and
(18) $\Phi_\nu = A_\nu$ on $\{x \in \mathbb{R}^D : \text{dist}(x, S_\nu) \geq 10^4\text{diam}(S_\nu)\}$ where A_ν is a proper Euclidean motion.

Next, for each ν ($1 \leq \nu \leq \nu_{\max}$), we pick a representative $y_\nu \in S_\nu$. Define

(19) $E = \{y_\nu : 1 \leq \nu \leq \nu_{\max}\}$.
(20) Thus $E \subset \mathbb{R}^d,\ 2 \leq \text{card}(E) \leq K$,
(21) $\frac{1}{2}\text{diam}(S) \leq \text{diam}(E) \leq \text{diam}(S)$ and by (12) and (13),
(22) $|x-y| \geq \tau \geq \exp(-C_K/\varepsilon)\text{diam}(S)$

for $x, y \in E$ distinct.

We prepare to apply a rescaled version of Theorem 11.26. For easier reading, let us note the assumptions and conclusions with the same notation there as we will need to verify and use them here.

- **Assumptions on E.**

  (23) $\text{card}(E) \leq K$
  (24) $|x-y| \geq \tau$ for $x, y \in E$ distinct.

- **Assumptions on ϕ.**

  (25) $(1+\delta)^{-1}|x-y| \leq |\phi(x)-\phi(y)| \leq (1+\delta)|x-y|,\ ,x,y \in E$.
  (26) ϕ has no negative η blocks.

- **Assumptions on the parameters.**

  (27) $0 < \eta < c\varepsilon\tau/\text{diam}(E)$ for small enough c
  (28) $C_K\delta^{1/\rho_K}\tau^{-1}\text{diam}(E) \leq \min(\varepsilon, \eta^d)$ for large enough C_K, ρ_K depending only on K and d.

- **Conclusion.**

  (28a) There exists a proper $C\varepsilon$-distorted diffeomorphism $\Psi : \mathbb{R}^d \to \mathbb{R}^d$ with the following properties:
  (29) $\Psi = \phi$ on E.

(30) Ψ agrees with a proper Euclidean motion on

$$\{x \in \mathbb{R}^d : \operatorname{dist}(x, E) \geq 1000\operatorname{diam}(E)\}.$$

(31) For each $z \in E$, Φ agrees with a proper Euclidean motion on $B(z, \tau/1000)$.

Let us check that our present $\phi : E \to \mathbb{R}^d$, $\delta, \varepsilon, \eta, \tau$ satisfy the hypotheses of Theorem 11.26. In fact: hypothesis (23) is (20). Hypothesis (24) is (22), Hypothesis (25) is immediate from (7). Hypothesis (26) is immediate from (8).

Let us check hypotheses (27) and (28). From (13) and (21) we have

(32) $\exp(-C_K/\varepsilon) \leq \tau/\operatorname{diam}(E)$.

Hence (27) and (28) will follow if we can show that the following two things:

(33) $0 < \eta < c\exp(-C_K/\varepsilon)$ for small enough c.

(34) $C_K\delta^{1/\rho_k}\exp(C_K/\varepsilon) \leq \min(\varepsilon, \eta^d)$. However, we now recall that δ and η are defined by (2) and (3). Thus (33) holds provided

(35) $L < c_K$ for small enough c_K and $L' > C_K$ for large enough C_K.

(36) Similarly (34) holds provided $L < c_K$ for small enough c_K and $1/\rho_K L'' - C_K \geq \max(1, dL')$. Assuming we can choose L, L', L'' as we wish, we have (33) and (34) hence also (27) and (28). This completes our verification of the hypothesis of Theorem 11.26 for our present Φ and E. Applying Theorem 11.26, we now obtain a proper $C\varepsilon$-distorted diffeomorphism $\Phi : \mathbb{R}^d \to \mathbb{R}^d$ satisfying (28a)–(31). For each $z \in E$, we now define a proper ε-distorted diffeomorphism Φ_z by setting:

(37) $\Phi_z = \Phi_\nu$ if $z = y_\nu$. (Recall (16), (19) and note that the y_ν, $1 \leq \nu \leq \nu_{\max}$ are distinct). From (17), (18), (37) we have the following:

(38) $\Phi_z = \phi$ on S_ν if $z = y_\nu$. In particular,

(39) $\Phi_z(z) = \phi(z)$ for each $z \in E$. Also

(40) $\Phi_z = A_z$ (a proper Euclidean motion) outside $B(z, 10^5\operatorname{diam}(S_\nu))$ if $z = y_\nu$. Recalling (11), we see that

(41) $\Phi_z = A_z$ (a proper Euclidean motion) outside $B(z, 10^5\exp(-5/\varepsilon)\tau)$. We prepare to apply Theorem 12.32, the gluing lemma to the present ϕ, E, $\Phi_z (z \in E)$, Ψ, ε and τ. Let us check the hypotheses of the gluing lemma. We have $\phi : E \to \mathbb{R}^d$ and $1/2|x - y| \leq |\phi(x) - \phi(y)| \leq 2|x - y|$ for $x, y \in E$ thanks to (7) provided

(41a) $L \leq 1$ and $L'' \geq 10$. See also (3). Also $|x - y| \geq \tau$ for $x, y \in E$ distinct, see (22). Moreover, for each $z \in E$, Φ_z is a proper ε distorted diffeomophism (see (16) and (37)). For each $z \in E$, we have $\Phi_z(z) = \phi(z)$ by (39) and $\Phi_z = A_z$ (a proper Euclidean motion) outside $B_1(z) = B(z, \exp(-4/\varepsilon)\tau)$, see (41). Here, we assume that,

(42) $L \leq c_k$ for a small enough c_k. Next, recall that Ψ satisfies (28a)–(31). Then Ψ is a $C\varepsilon$-distorted diffeomorphism, $\Psi = \Phi$ on E and for $z \in E$,

Ψ agrees with a proper Euclidean motion A_z^* on $B(z, \frac{\tau}{1000})$, hence on $B_4(z) = B(z, \exp(-1/\varepsilon)\tau)$. Here, again, we assume that L satisfies (42). This completes the verification of the hypotheses of the gluing lemma. Applying that lemma, we obtain:

(43) a $C'\varepsilon$-distorted diffeomorphism $\Phi : \mathbb{R}^d \to \mathbb{R}^d$ such that

(44) $\Phi = \Phi_z$ on $B_2(z) = B(z, \exp(-3/\varepsilon)\tau)$, for each $z \in E$ and

(45) $\Phi = \Psi$ outside $\cup_{z\in E} B_3(z) = \cup_{z\in E} B(z, \exp(-2/\varepsilon)\tau)$. Since Ψ is proper, we know that

(46) Φ is proper. Let $z = y_\mu \in E$. Then (11) shows that $S_\mu \subset B(z, \exp(-5/\varepsilon)\tau)$ and, therefore, (44) yields $\Phi = \Phi_z$ on S_μ for $z = y_\mu$. Together, with (38), this yields $\Psi = \phi$ on S_μ for each $\mu(1 \le \mu \le \mu_{\max})$. Since the $S_\mu(1 \le \mu \le \mu_{\max})$ form a partition of S, we conclude that

(47) $\Phi = \phi$ on S. Moreover, suppose that

$$\operatorname{dist}(x, S) \ge 10^4 \operatorname{diam}(S).$$

Then x does not belong to $B(z, \exp(-2/\varepsilon)\tau)$ for any $z \in E$ as we see from (13). (Recall that $E \subset S$). Consequently (45) yields $\Phi(x) = \Psi(x)$ and, therefore, (30) tells us that $\Psi(x) = A_\infty(x)$. Since $\operatorname{dist}(x, S) \ge 10^4\operatorname{diam}(S)$, we have

$$\operatorname{dist}(x, E) \ge \operatorname{dist}(x, S) \ge 10^4\operatorname{diam}(S) \ge 10^3 \operatorname{diam} E.$$

Hence (30) applies. Thus,

(48) Φ agrees with a proper Euclidean motion A_∞^* on

$$\{x \in \mathbb{R}^d : \operatorname{dist}(x, S) \ge 10^4\operatorname{diam}(S)\}.$$

Collecting our results (43), (46), (47), (48), we have the following:

(49) There exists a proper $C\varepsilon$-distorted diffeomorphism Φ such that $\Phi = \phi$ on S and Φ agrees with a proper Euclidean motion on

$$\{x \in \mathbb{R}^d : \operatorname{dist}(x, S) \ge 10^4\operatorname{diam}(S)\}.$$

We have established (49) assuming that the small constant L and the large constants L', L'' satisfy the conditions (14), (35), (36), (41a), (42). By picking L first, L' second and L'' third, we can satisfy all those conditions with $L = c_K$, $L' = C_K'$, $L'' = C_K''$. With these L, L', L'' we have shown that (1)–(8) together imply that (49) holds. Thus, we have proven the following:

(50) For suitable constants C, C_K, C_K', C_K'' depending only on d and K the following holds: suppose that $0 < \varepsilon < c_k$. Set $\eta = \exp(-C_K'/\varepsilon)$ and $\delta = \exp(-C_K''/\varepsilon)$. Let $\phi : S \to \mathbb{R}^d$ with $\operatorname{card}(S) = K$, $S \subset \mathbb{R}^d$. Assume that

$$(1+\delta)^{-1}|x-y| \le |\phi(x) - \phi(y)| \le (1+\delta)|x-y|,\ x, y \in S.$$

Then if ϕ has no negative η block, then there exists a proper $C\varepsilon$-distorted diffeomorphism $\Phi : \mathbb{R}^d \to \mathbb{R}^d$ such that $\phi = \Phi$ on S and Φ agrees with a proper Euclidean motion on

$$\{x \in \mathbb{R}^d : \text{dist}(x, S) \geq 10^4 \text{diam}(S)\}.$$

Taking ε to be ε/C, we thus deduce:

(51) For suitable constants C_{new}, C'_{new}, C''_{new} depending only on d and K the following holds: suppose that $0 < \varepsilon < c_{\text{new}}$. Set $\eta = \exp(-C'_{\text{new}}/\varepsilon)$ and $\delta = \exp(-C''_{\text{new}}/\varepsilon)$. Let $\phi : S \to \mathbb{R}^d$ with card$(S) = K$, $S \subset \mathbb{R}^d$. Assume that

$$(1+\delta)^{-1}|x-y| \leq |\phi(x) - \phi(y)| \leq (1+\delta)|x-y|,\ x, y \in S.$$

Then if ϕ has no negative η block, then there exists a proper ε-distorted diffeomorphism $\Phi : \mathbb{R}^d \to \mathbb{R}^d$ such that $\phi = \Phi$ on S and Φ agrees with a proper Euclidean motion on

$$\{x \in \mathbb{R}^d : \text{dist}(x, S) \geq 10^4 \text{diam}(S)\}.$$

That is almost Theorem 11.27 except we are assuming card$(S) = K$ rather than card$(S) \leq K$. Therefore we proceed as follows: we have our result (50) and we have an inductive hypothesis. We now take $C' = \max(C'_{\text{old}}, C'_{\text{new}})$, $C'' = \text{old}(C''_{\text{max}}, C''_{\text{new}})$ and $c' = \min(c'_{\text{old}}, c'_{\text{new}})$. These constants are determined by d and K. We now refer to $\eta_{\text{old}}, \eta_{\text{new}}, \eta, \delta_{\text{old}}, \delta_{\text{new}}, \delta$ to denote $\exp(-C'_{\text{old}}/\varepsilon)$, $\exp(-C'_{\text{new}}/\varepsilon)$, $\exp(-C'/\varepsilon)$, $\exp(-C''_{\text{old}}/\varepsilon)$, $\exp(-C''_{\text{new}}/\varepsilon)$, $\exp(-C''/\varepsilon)$ respectively.

Note that $\delta \leq \delta_{\text{old}}$, $\delta \leq \delta_{\text{new}}$, $\eta \leq \eta_{\text{old}}$ and $\eta \leq \eta_{\text{new}}$. Also if $0 < \varepsilon < c$, then $0 < \varepsilon < c_{\text{old}}$ and $0 < \varepsilon < c_{\text{new}}$. If

$$(1+\delta)^{-1}|x-y| \leq |\phi(x) - \phi(y)| \leq (1+\delta)|x-y|, x, y \in S$$

then the same holds for δ_{old} and δ_{new}. Also if ϕ has no negative η-block, then it has no negative η_{old}-block and it has no negative η_{new}-block. Consequently by using (51) and the induction hypothesis, we have proved Theorem 11.27. □

We now give

The Proof of Theorem 11.28 To see this, we simply observe that increasing C''_K in Theorem 11.27 merely weakens the result so we may increase C''_K and achieve that $0 < \delta < c\eta^d$ for small enough c and $0 < \delta < \varepsilon$. The desired result then follows by using the fact that if $\phi : S \to R^d$ extends to a δ-distorted diffeomorphism and ϕ has a positive η block (respectively negative η block) for $0 < \delta < c\eta^d$ for small enough c, then ϕ is proper (respectively improper). □

12.5 Proofs of Theorem 11.31, Theorem 11.30 and Theorem 11.29

The Proof of Theorem 11.31 and Theorem 11.30 Pick c_K, C_K, C''_K as in Theorem 11.27 and Theorem 11.28. Let δ and η be as in Theorem 11.27 and let us take $S_0 = \{x, y\}$. Then we see that

$$(1 + \delta)|x - y| \leq |\phi(x) - \phi(y)| \leq (1 + \delta)|x - y|, x, y \in S.$$

Now, if ϕ has no negative η-block, then by Theorem 11.27, Φ exists with the properties claimed. Similarly, if ϕ has no positive η-block, then applying Theorem 11.27 to the function ϕo(reflection) we obtain the Φ we need with the properties claimed. Suppose that ϕ has a positive η-block $(x_0, ...x_D)$ and a negative η-block $(y_0, ...y_D)$. Then $\phi|_{\{x_0,...,x_D,y_0,...,y_D\}}$ cannot be extended to a δ distorted diffeomorphism $\Phi : \mathbb{R}^d \to \mathbb{R}^d$. Indeed, the η-block $(x_0, ..., x_d)$ forces any such Φ to be proper while the η-block $(y_0, ...y_d)$ forces Φ to be improper. Since card $\{x_0, ..., x_d, y_0, ...y_d\} \leq 2(d+1)$, the proofs of Theorem 11.31 and Theorem 11.30 are complete. □

The Proof of Theorem 11.29 Take $k \leq d + 1$ and apply Theorem 11.27. Let η and δ be determined by ε as in Theorem 11.27. If ϕ has no negative η-block then applying Theorem 11.27 to ϕ or ϕo(reflection), we see that ϕ extends to a ε-distorted diffeomorphism $\Phi : \mathbb{R}^d \to \mathbb{R}^d$. However, since (card)$(S) \leq d + 1$, the only possible (negative or positive) η-block for ϕ is all of S. Thus either ϕ has a negative η-block or it has no positive η-block. □

13

Extensions of Smooth Small Distortions [41]: Introduction

In Chapter 8, using algebraic geometry, we discovered a quantitative relationship: $\delta = c\varepsilon^{c'}$. Here the points in the set E are forced to live on an ellipse. In Chapters 11–12, we discovered that when the points of the set E live close to a hyperplane, are not too close to each other, and have not too large a diameter then $\delta = \exp\left(-\frac{C_K}{\varepsilon}\right)$ where the constant C_K depends on d and the constant K is chosen at the same time as d. In this chapter, [41], we will work in $\mathbb{R}^n$ and achieve the optimal $c\delta = \varepsilon$, c depending on n.

13.1 Class of Sets *E*

We work with a C^1 map $\phi : U \to \mathbb{R}^n$ with $U \subset \mathbb{R}^n$ an open set. For a compact set E in $\mathbb{R}^n$, and $x \in \mathbb{R}^n$, we write for convenience $d(x) := \text{dist}(x, E)$. Let $\varepsilon > 0$.

We assume the following:

(a) Geometry of E. For certain positive constants c_0, C_1, c_2 depending on n, the following holds: let $x \in \mathbb{R}^n \backslash E$. If $d(x) \leq c_0\, \text{diam}(E)$, then there exists a ball $B(z, r) \subset E$ such that $|z - x| \leq C_1\, d(x)$ and $r \geq c_2\, d(x)$.

(b) Geometry of ϕ. For $x, y \in E$, $|x - y|(1 - \varepsilon) \leq |\phi(x) - \phi(y)| \leq (1 + \varepsilon)|x - y|$.

(c) Controlled constants: ε is smaller than a small positive constant determined by c_0, C_1, c_2, n.

Here is our main result of this chapter.

13.2 Main Result

Theorem 13.34. *Under the above assumptions, there exists a C^1 function $\Phi : \mathbb{R}^n \to \mathbb{R}^n$ and a Euclidean motion $A : \mathbb{R}^n \to \mathbb{R}^n$, with the following properties:*

Near Extensions and Alignment of Data in $\mathbb{R}^n$: Whitney extensions of near isometries, shortest paths, equidistribution, clustering and non-rigid alignment of data in Euclidean space,
First Edition. Steven B. Damelin.

- $(1-c'\varepsilon)\,|x-y| \leq |\Phi(x)-\Phi(y)| \leq (1+c'\varepsilon)\,|x-y|$ *for all* $x, y \in \mathbb{R}^n$. c' *determined by* c_0, C_1, c_2, n.
- $\Phi = \phi$ *in a neighborhood of* E.
- $\Phi = A$ *outside* $\{x \in \mathbb{R}^n : d(x) < c_0\}$.
- $\Phi : \mathbb{R}^n \to \mathbb{R}^n$ *is one-to-one and onto.*
- *If* $\phi \in C^m(U)$ *for some given* $m \geq 1$, *then* $\Phi \in C^m(\mathbb{R}^n)$.
- *If* $\phi \in C^\infty(U)$, *then* $\Phi \in C^\infty(\mathbb{R}^n)$.

We now proceed to prove Theorem 13.34.

14

Extensions of Smooth Small Distortions: First Results

Lemma 14.1. *Let $B(z, r) \subset E$. Then there exists a Euclidean motion A, such that for every $L \geq 1$, and for every*

$$y \in B(z, Lr) \cap E, \text{ we have } |\phi(y) - A(y)| \leq CL^2 \varepsilon r. \tag{14.1}$$

Proof. Without loss of generality, we may assume $z = 0$, $r = 1$, and $\phi(0) = 0$. Let $e_1, \ldots, e_n$ be the unit vector in $\mathbb{R}^n$. We have $1 - \varepsilon \leq |\phi(e_i)| \leq 1 + \varepsilon$ for each i, and $(1 - \varepsilon)\sqrt{2} \leq |\phi(e_i) - \phi(e_j)| \leq (1 + \varepsilon)\sqrt{2}$ for $i \neq j$. Since $-2\phi(e_i) \cdot \phi(e_j) = |\phi(e_i) - \phi(e_j)|^2 - |\phi(e_i)|^2 - |\phi(e_j)|^2$, it follows that

$$|\phi(e_i) \cdot \phi(e_j) - \delta_{ij}| \leq C\varepsilon \text{ for each } i, j,$$
$$\text{where } \delta_{ij} \text{ is the Kronecker delta function.} \tag{14.2}$$

Let $A \in \mathcal{O}(n)$ be the orthogonal matrix whose columns arise by applying the Gram–Schmidt process to the vectors $\phi(e_1)$, $\phi(e_2)$, ..., $\phi(e_n)$. Then (14.2) implies the estimate

$$|\phi(e_i) - Ae_i| \leq C\varepsilon \text{ for each } i.$$

Replacing ϕ by $A^{-1} \circ \phi$, we may therefore assume without loss of generality that

$$|\phi(e_i) - e_i| \leq C\varepsilon \text{ for each } i. \tag{14.3}$$

Assume (14.3), and recalling that $\phi(0) = 0$, we will prove (14.1) with $A = I$. Thus, let $L \geq 1$, and let $y \in B(0, L) \cap E$. We have $(1 - \varepsilon)\,|y| \leq |\phi(y)| \leq (1 + \varepsilon)\,|y|$, hence

$$\big||\phi(y)| - |y|\big| \leq \varepsilon L. \tag{14.4}$$

In particular,

$$|\phi(y)| \leq \varepsilon L. \tag{14.5}$$

Near Extensions and Alignment of Data in $\mathbb{R}^n$: Whitney extensions of near isometries, shortest paths, equidistribution, clustering and non-rigid alignment of data in Euclidean space,
First Edition. Steven B. Damelin.

We have

$$(1-\varepsilon)\,|y-e_i| \leq |\phi(y)-\phi(e_i)| \leq (1+\varepsilon)\,|y-e_i| \text{ for each } i.$$

Hence, by (14.3) and (14.5), we have

$$\Big||\phi(y)-e_i|-|y-e_i|\Big| \leq C\varepsilon L \text{ for each } i. \tag{14.6}$$

From (14.4), (14.5), (14.6), we see that

$$\Big||\phi(y)|^2-|y|^2\Big| = \Big(|\phi(y)|+|y|\Big)\cdot\Big||\phi(y)|-|y|\Big| \leq CL^2\varepsilon, \tag{14.7}$$

and similarly,

$$\Big||\phi(y)-e_i|^2-|y-e_i|^2\Big| \leq CL^2\varepsilon. \tag{14.8}$$

Since

$$\begin{aligned} -2\phi(y)\cdot e_i &= |\phi(y)-e_i|^2-|\phi(y)|^2-1 \text{ and} \\ -2y\cdot e_i &= |y-e_i|^2-|y|^2-1, \end{aligned}$$

it follows from (14.7) and (14.8) that

$$\Big|[\phi(y)-y]\cdot e_i\Big| \leq CL^2\varepsilon \text{ for each } i.$$

Consequently, $|\phi(y)-y| \leq CL^2 e$, proving (14.1) and A = identity. ∎

When we apply Lemma 14.1, we will always take L to be a controlled constant.

The proof of the following Lemma is straightforward, and may be left to the reader. Note that $\nabla A(x)$ is independent of $x \in \mathbb{R}^n$ when $A : \mathbb{R}^n \to \mathbb{R}^n$ is an affine function. We write ∇A in this case without indicating x.

Lemma 14.2. *Let $B(z,r)$ be a ball, let $A : \mathbb{R}^n \to \mathbb{R}^n$ be an affine function, and let $M > 0$ be a real number. If $|A(y)| \leq M$ for all $y \in B(z,r)$, then $|\nabla A| \leq CM/r$, and for any $L \geq 1$ and $y \in B(z,Lr)$ we have*

$$\Big|A(y)\Big| \leq CLM.$$

When we apply Lemma 14.2, we will always take L to be a controlled constant.

Lemma 14.3. *For $\eta > 0$ small enough, we have*

$$(1-C\varepsilon)I \leq \Big(\nabla\phi(y)\Big)^{+}\Big(\nabla\phi(y)\Big) \leq (1+C\varepsilon)I \textit{ for all } y \in \mathbb{R}^n \textit{ s.t. } d(y)<\eta. \tag{14.9}$$

Proof. If y is an interior point of E, then we have (14.9). Suppose y is a boundary point of E. Arbitrarily close to y, we can find $x \in \mathbb{R}^n\backslash E$. We have an interior point

z in E such that $|z - x| \leq C_1 d(x) \leq C_1|y - x|$, hence $|z - y| \leq (1 + C_1)|y - x|$. Since z is an interior point of E, we have

$$(1 - C\varepsilon)I \leq \left(\nabla\phi(z)\right)^{+}\left(\nabla\phi(z)\right) \leq (1 + C\varepsilon), \tag{14.10}$$

as observed above. However, we can make $|z-y|$ as small as we like here, simply by taking $|y-x|$ small enough. Since $\phi \in C^1(U)$, we may pass to the limit, and deduce (14.9) from (14.10). Thus, (14.9) holds for all $y \in E$. Since $E \subset U$ is compact and $\phi \in C^1(U)$, the lemma now follows. ■

Lemma 14.4. *For $\eta > 0$ small enough, we have*

$$\left|\phi(y) - \left[\phi(x) + \nabla\phi(x) \cdot (y - x)\right]\right| \leq \varepsilon\left|y - x\right|$$

for all $x, y \in U$ such that $d(x) \leq \eta$ and $|y - x| \leq \eta$.

Proof. If η is small enough and $d(x) \leq \eta$, then $B(x, \eta) \subset U$ and $|\nabla\phi(y) - \nabla\phi(x)| \leq \varepsilon$ for all $y \in B(x, \eta)$. (These remarks follow from the fact that $E \subset U$ is compact and $\phi \in C^1(U)$.)

The lemma now follows from the fundamental theorem of calculus. ■

Lemma 14.5. *Let $\Psi : \mathbb{R}^n \to \mathbb{R}^n$ be a C^1 function. Assume that* $\det \nabla\Psi \neq 0$ *everywhere on $\mathbb{R}^n$, and assume that Ψ agrees with a Euclidean motion outside a ball B. Then $\Psi : \mathbb{R}^n \to \mathbb{R}^n$ is one-to-one and onto.*

Proof. Without loss of generality, we may suppose $\Psi(x) = x$ for $|x| \geq 1$. First we show that Ψ is onto. Since $\det \nabla\Psi \neq 0$, we know that $\Psi(\mathbb{R}^n)$ is open, and of course $\Psi(\mathbb{R}^n)$ is non-empty. If we can show that $\Psi(\mathbb{R}^n)$ is closed, then it follows that $\Psi(\mathbb{R}^n) = \mathbb{R}^n$, i.e., Ψ is onto.

Let $\{x_\nu\}_{\nu \geq 1}$ be a sequence converging to $x_\infty \in \mathbb{R}^n$, with each $x_\nu \in \Psi(\mathbb{R}^n)$. We show that $x_\infty \in \Psi(\mathbb{R}^n)$. Let $x_\nu = \Psi(y_\nu)$. If infinitely many y_ν satisfy $|y_\nu| \geq 1$, then infinitely many x_ν satisfy $|x_\nu| \geq 1$, since $x_\nu = \Psi(y_\nu) = y_\nu$ for $|y_\nu| \geq 1$. Hence, $|x_\infty| \geq 1$ in this case, and consequently

$$x_\infty = \Psi(x_\infty) \in \Psi(\mathbb{R}^n).$$

On the other hand, if only finitely many y_ν satisfy $|y_\nu| \geq 1$, then there exists a convergent subsequence $y_{\nu_i} \to y_\infty$ as $i \to \infty$. In this case, we have

$$x_\infty = \lim_{i\to\infty} \Psi(y_{\nu_i}) = \Psi(y_\infty) \in \Psi(\mathbb{R}^n).$$

Thus, in all cases, $x_\infty \in \Psi(\mathbb{R}^n)$. This proves that $\Psi(\mathbb{R}^n)$ is closed, and therefore $\Psi : \mathbb{R}^n \to \mathbb{R}^n$ is onto.

Let us show that Ψ is one-to-one. We know that Ψ is bounded on the unit ball. Fix M such that

$$\left|\Psi(y)\right| \leq M \text{ for } |y| \leq 1. \tag{14.11}$$

We are assuming that $\Psi(y) = y$ for $|y| \geq 1$. For $|x| > \max(M, 1)$, it follows that $y = x$ is the only point $y \in \mathbb{R}^n$ such that $\Psi(y) = x$. Now let $Y = \{y' \in \mathbb{R}^n : \Psi(y') = \Psi(y'')$ for some $y'' \neq y'\}$. The set Y is bounded, thanks to (14.11). Also, the inverse function theorem shows that Y is open. We will show that Y is closed. This implies that Y is empty, proving that $\Psi : \mathbb{R}^n \to \mathbb{R}^n$ is one-to-one.

Thus, let $\{y'_\nu\}_{\nu \geq 1}$ be a convergent sequence, with each $y'_\nu \in Y$; suppose $y'_\nu \to y'_\infty$ as $\nu \to \infty$. We will prove that $y'_\infty \in Y$.

For each ν, pick $y''_\nu \neq y'_\nu$ such that

$$\Psi(y''_\nu) = \Psi(y'_\nu).$$

Each y''_ν satisfies $|y''_\nu| \leq \max(M.1)$, thanks to (14.11).

Hence, after passing to a subsequence, we may assume $y''_\nu \to y''_\infty$ as $\nu \to \infty$. We already know that $y'_\nu \to y'_\infty$ as $\nu \to \infty$.

Suppose $y'_\infty = y''_\infty$. Then arbitrarily near y'_∞ there exist pairs y'_ν, y''_ν, with $y'_\nu \neq y''_\nu$ and $\Psi(y'_\nu) = \Psi(y''_\nu)$. This contradicts the inverse function theorem, since $\det \nabla\Psi(y'_\infty) \neq 0$.

Consequently, we must have $y'_\infty \neq y''_\infty$. Recalling that $\Psi(y'_\nu) = \Psi(y''_\nu)$, and passing to the limit, we see that $\Psi(y'_\infty) = \Psi(y''_\infty)$.

By definition, we therefore have $y'_\infty \in Y$, proving that Y is closed, as asserted above. Hence, Y is empty, and $\Psi : \mathbb{R}^n \to \mathbb{R}^n$ is one-to-one. ■

From now on, we assume without loss of generality that

$$\operatorname{diam} E = 1. \tag{14.12}$$

15

Extensions of Smooth Small Distortions: Cubes, Partitions of Unity, Whitney Machinery

15.1 Cubes

$\mathbb{R}^n \backslash E$ is partitioned into "cubes" $\{Q_\nu\}$. We write β_ν to denote the sidelength of Q_ν, and we write Q_ν^* to denote the cube Q_ν, dilated about its center by a factor of 3. The cubes have the following properties,

$$c\beta_\nu \leq d(x) \leq C\beta_\nu \text{ for all } x \in Q_\nu^*. \tag{15.1}$$

$$\text{Any given } x \in \mathbb{R}^n \text{ belongs to } Q_\nu^* \text{ for at most } C \text{ distinct } \nu. \tag{15.2}$$

15.2 Partition of Unity

For each Q_ν, we have a cutoff function $\Theta_\nu \in C^\infty(\mathbb{R}^n)$, with the following properties,

$$\Theta_\nu \geq 0 \text{ on } \mathbb{R}^n. \tag{15.3}$$

$$\text{supp}\,\Theta_\nu \subset Q_\nu^*. \tag{15.4}$$

$$|\nabla\Theta_\nu| \leq C\beta_\nu^{-1} \text{ on } \mathbb{R}^n. \tag{15.5}$$

$$\sum_\nu \Theta_\nu = 1 \text{ on } \mathbb{R}^n \backslash E. \tag{15.6}$$

15.3 Regularized Distance

A function $\delta(x)$, defined on $\mathbb{R}^n$, has the following properties,

$$cd(x) \leq \delta(x) \leq Cd(x) \text{ for all } x \in \mathbb{R}^n. \tag{15.7}$$

Near Extensions and Alignment of Data in $\mathbb{R}^n$: Whitney extensions of near isometries, shortest paths, equidistribution, clustering and non-rigid alignment of data in Euclidean space,
First Edition. Steven B. Damelin.

$$\delta(\cdot) \text{ belongs to } C^{\infty}_{\text{loc}}(\mathbb{R}^n \setminus E). \tag{15.8}$$

$$|\nabla \delta(x)| \leq C \text{ for all } x \in \mathbb{R}^n \setminus E. \tag{15.9}$$

Thanks to (15.1) and (15.7), the following holds,

$$\left[\begin{array}{l} \text{Let } x \in \mathbb{R}^n, \text{ and let } Q_\nu \text{ be one of the cubes.} \\ \text{If } d(x) \geq c_0 \text{ and } x \in Q^*_\nu, \text{ then } \beta_\nu > c_3. \end{array} \right. \tag{15.10}$$

Recall that diam $E = 1$.

Let Q_ν be a cube such that $\beta_\nu \leq c_3$. Then $d(x) < c_0$ for all $x \in Q^*_\nu$, as we see from (15.10). Let x_ν be the center of Q_ν. Since $d(x_\nu) < c_0$, take $x = x_\nu$ and obtain a ball

$$B(z_\nu, r_\nu) \subset E, \tag{15.11}$$

such that

$$c d(x_\nu) < r_\nu \leq C d(x_\nu), \tag{15.12}$$

and

$$|z_\nu - x_\nu| \leq C d(x_\nu). \tag{15.13}$$

The ball $B(z_\nu, r_\nu)$ has been defined whenever $\beta_\nu \leq c_3$. (To see that $r_\nu \leq C d(x_\nu)$, we just note that $B(z_\nu, r_\nu) \subset E$ but $x_\nu \notin E$; hence $|z_\nu - x_\nu| > r_\nu$, and therefore (15.13) implies $r_\nu \leq C d(x_\nu)$.)

From (15.12), (15.13) and (15.1), (15.7), we learn the following,

$$Q^*_\nu \subset B(z_\nu, C r_\nu). \tag{15.14}$$

$$c\delta(x) < r_\nu < C\delta(x) \text{ for any } x \in Q^*_\nu. \tag{15.15}$$

$$|z_\nu - x| \leq C\delta(x) \text{ for any } x \in Q^*_\nu. \tag{15.16}$$

These results (and (15.11)) in turn imply the following,

$$\text{Let } x \in Q^*_\mu \cap Q^*_\nu. \text{ Then } B(z_\nu, r_\nu) \subset B(z_\mu, C r_\mu) \cap E. \tag{15.17}$$

Here, (15.14), (15.15), (15.16) hold whenever $\beta_\nu \leq c_3$; while (15.17) holds whenever $\beta_\mu, \beta_\nu \leq c_3$.

We want an analogue of $B(z_\nu, r_\nu)$ for cubes Q_ν such that $\beta_\nu > c_3$.

There exists $x \in \mathbb{R}^n$ such that $d(x) = c_0/2$. Using this x, we obtain a ball

$$B(z_\infty, r_\infty) \subset E, \tag{15.18}$$

such that

$$c < r_\infty \leq 1/2. \tag{15.19}$$

(We have $r_\infty \leq 1/2$, simply because diam $E = 1$.)

From (15.18), (15.19) and the fact that diam $E = 1$, we conclude that

$$E \subset B(z_\infty, Cr_\infty). \tag{15.20}$$

16

Extensions of Smooth Small Distortions: Picking Motions

For each cube Q_ν, we pick a Euclidean motion A_ν, as follows,

Case I ("Small" Q_ν.) Suppose $\beta_\nu \leq c_3$. Applying Lemma 14.1 to the ball $B(z_\nu, r_\nu)$, we obtain a Euclidean motion A_ν with the following property.

$$\text{For } L \geq 1 \text{ and } y \in B(z_\nu, Lr_\nu) \cap E, \text{ we have } \left|\phi(y) - A_\nu(y)\right| \leq CL^2 \varepsilon r_\nu. \quad (16.1)$$

Case II ("Not-so-small" Q_ν.) Suppose $\beta_\nu > c_3$. Applying Lemma 14.1 to the ball $B(z_\infty, r_\infty)$, we obtain a Euclidean motion A_∞ with the following property.

$$\text{For } L \geq 1 \text{ and } y \in B(z_\infty, Lr_\infty) \cap E, \text{ we have } \left|\phi(y) - A_\infty(y)\right| \leq CL^2 \varepsilon r_\infty. \quad (16.2)$$

In case II, we define

$$A_\nu = A_\infty. \quad (16.3)$$

Thus, $A_\nu = A_{\nu'}$ whenever ν and ν' both fall into Case II. Note that (16.2) together with (15.19) and (15.20) yield the estimate

$$\left|\phi(y) - A_\infty(y)\right| \leq C\varepsilon \text{ for all } y \in E. \quad (16.4)$$

The next result establishes the mutual consistency of the A_ν.

Lemma 16.1. *For $x \in Q_\mu^* \cap Q_\nu^*$, we have*

$$\left|A_\mu(x) - A_\nu(x)\right| \leq C\varepsilon\delta(x), \quad (16.5)$$

and

$$\left|\nabla A_\mu - \nabla A_\nu\right| \leq C\varepsilon. \quad (16.6)$$

Near Extensions and Alignment of Data in $\mathbb{R}^n$: Whitney extensions of near isometries, shortest paths, equidistribution, clustering and non-rigid alignment of data in Euclidean space,
First Edition. Steven B. Damelin.

Proof. We proceed by cases.

Case 1: Suppose $\beta_\mu, \beta_\nu \leq c_3$. Then A_ν satisfies (16.1), and A_μ satisfies the analogous condition for $B(z_\mu, r_\mu)$. Recalling (15.17), we conclude that

$$\left|\phi(y) - A_\mu(y)\right| \leq C\varepsilon r_\mu \text{ for } y \in B(z_\mu, r_\mu), \tag{16.7}$$

and

$$\left|\phi(y) - A_\nu(y)\right| \leq C\varepsilon r_\nu \text{ for } y \in B(z_\nu, r_\nu). \tag{16.8}$$

Moreover, (15.15) gives

$$c\delta(x) < r_\mu < C\delta(x) \text{ and } c\delta(x) < r_\nu < C\delta(x). \tag{16.9}$$

By (16.7), (16.8), (16.9), we have

$$\left|A_\mu(y) - A_\nu(y)\right| \leq C\varepsilon r_\nu \text{ for } y \in B(z_\nu, r_\nu). \tag{16.10}$$

Now, $A_\mu(y) - A_\nu(y)$ is an affine function. Hence, Lemma 14.2 and inclusion (15.14) allows us to deduce from (16.10) that

$$\left|A_\mu(y) - A_\nu(y)\right| \leq C\varepsilon r_\nu \text{ for all } y \in Q_\nu^*, \tag{16.11}$$

and

$$\left|\nabla A_\mu - \nabla A_\nu\right| \leq C\varepsilon. \tag{16.12}$$

Since $x \in Q_\nu^*$, the desired estimates (16.5), (16.6) follow at once from (16.9), (16.11) and (16.12). Thus, Lemma 16.1 holds in Case 1.

Case 2: Suppose $\beta_\nu \leq c_3$ and $\beta_\mu > c_3$. Then by (16.1) and (15.11), A_ν satisfies

$$\left|\phi(y) - A_\nu(y)\right| \leq C\varepsilon r_\nu \text{ for } y \in B(z_\nu, r_\nu); \tag{16.13}$$

whereas $A_\mu = A_\infty$, so that (16.4) and (15.11) give

$$\left|\phi(y) - A_\mu(y)\right| \leq C\varepsilon \text{ for all } y \in B(z_\nu, r_\nu). \tag{16.14}$$

Since $x \in Q_\mu^* \cap Q_\nu^*$, (15.1) and (15.7) give

$$c\delta(x) \leq \beta_\mu \leq C\delta(x) \text{ and } c\delta(x) \leq \beta_\nu \leq C\delta(x).$$

In this case, we also have $\beta_\nu \leq c_3$ and $\beta_\mu > c_3$. Consequently,

$$c < \beta_\mu < C, c < \beta_\nu < C, \text{ and } c < \delta(x) < C. \tag{16.15}$$

By (15.15), we also have

$$c < r_\nu < C. \tag{16.16}$$

From (16.13), (16.14), (16.16), we see that

$$\left|A_\mu(y) - A_\nu(y)\right| \leq C\varepsilon \text{ for all } y \in B_\nu(z_\nu, r_\nu). \tag{16.17}$$

Lemma 14.2, estimate (16.16) and inclusion (16.14) let us deduce from (16.17) that

$$\left|A_\mu(y) - A_\nu(y)\right| \leq C\varepsilon \text{ for all } y \in Q_\nu^*, \tag{16.18}$$

and

$$\left|\nabla A_\mu - \nabla A_\nu\right| \leq C\varepsilon. \tag{16.19}$$

Since $x \in Q_\nu^*$, the desired estimates (16.5), (16.6) follow at once from (16.15), (16.18), (16.19). Thus, Lemma 16.1 holds in Case 2.

Case 3: Suppose $\beta_\nu > c_3$ and $\beta\mu \leq c_3$. Reversing the roles of Q_μ and Q_ν, we reduce matters to Case 2. Thus, Lemma 16.1 holds in Case 3.

Case 4: Suppose $\beta_\mu, \beta_\nu > c_3$. Then by definition $A_\mu = A_\nu = A_\infty$, and estimates (16.5), (16.6) hold trivially. Thus, Lemma 16.1 holds in Case 4.

We have proved the desired estimates (16.5), (16.6) in all cases. ∎

The following lemma shows that A_ν closely approximates ϕ on Q_ν^* when Q_ν^* lies very close to E.

Lemma 16.2. *For $\eta > 0$ small enough, the following holds.*
Let $x \in Q_\nu^*$, and suppose $\delta(x) \leq \eta$. Then $x \in U$, $|\phi(x) - A_\nu(x)| \leq C\varepsilon\delta(x)$, and $|\nabla\phi(x) - \nabla A_\nu| \leq C\varepsilon$.

Proof. We have $\beta_\nu < C\delta(x) \leq C\eta$ by (15.1) and (15.7). If η is small enough, it follows that $\beta_\nu < c_3$, so Q_ν falls into Case I, and we have

$$\left|\phi(y) - A_\nu(y)\right| \leq C\varepsilon r_\nu \text{ for } y \in B(z_\nu, r_\nu) \tag{16.20}$$

by (16.1). Also, (15.15), (15.16) show that

$$B(z_\nu, r_\nu) \subset B\big(x, C\delta(x)\big) \subset B(x, C\eta). \tag{16.21}$$

We have

$$d(x) \leq C\delta(x) \leq C\eta \tag{16.22}$$

by (15.7). (In particular, $x \in U$ if η is small enough.) If η is small enough, then (16.21), (16.22) and Lemma 14.4 imply

$$y \in U \text{ and } \left|\phi(y) - \left[\phi(x) + \nabla\phi(x)\cdot(y - x)\right]\right| < \varepsilon\left|y - x\right| \text{ for } y \in B(z_\nu, r).$$

Hence, by (16.21) and (15.15), we obtain the estimate

$$\left|\phi(y) - [\phi(x) + \nabla\phi(x) \cdot (y - x)]\right| \leq C\varepsilon r_\nu \text{ for } y \in B(z_\nu, r_\nu). \tag{16.23}$$

Combining (16.20) with (16.23), we find that

$$\left|A_\nu(y) - [\phi(x) + \nabla\phi(x) \cdot (y - x)]\right| \leq C\varepsilon r_\nu \text{ for } y \in B(z, r_\nu). \tag{16.24}$$

The function $y \to A_\nu(y) - [\phi(x) + \nabla\phi(x) \cdot (y - x)]$ is affine. Hence, estimate (16.24), inclusion (15.14), and Lemma 14.2 together tell us that

$$\left|A_\nu(y) - [\phi(x) + \nabla\phi(x) \cdot (y - x)]\right| \leq C\varepsilon r_\nu \text{ for } y \in Q_\nu^*, \tag{16.25}$$

and

$$\left|\nabla A_\nu - \nabla\phi(x)\right| \leq C\varepsilon. \tag{16.26}$$

Since $x \in Q_\nu^*$, we learn from (16.25) and (15.15) that

$$\left|A_\nu(x) - \phi(x)\right| \leq C\varepsilon\delta(x). \tag{16.27}$$

Estimates (16.26) and (16.27) (and an observation that $x \in U$) are the conclusions of Lemma 16.2. ■

17

Extensions of Smooth Small Distortions: Unity Partitions

Our plan is to patch together the function ϕ and the Euclidean motion A_ν, using a partition of unity on $\mathbb{R}^n$. Note that the Θ_ν in Section 15 sums to 1 only on $\mathbb{R}^n \setminus E$.

Let $\eta > 0$ be a small enough number. Let $\chi(t)$ be a C^∞ function on $\mathbb{R}$, having the following properties

$$\begin{cases} 0 \le \chi(t) \le 1 & \text{for all } t; \\ \chi(t) = 1 & \text{for } t \le \eta; \\ \chi(t) = 0 & \text{for } t \ge 2\eta; \\ |\chi'(t)| \le C\eta^{-1} & \text{for all } t. \end{cases} \tag{17.1}$$

We define

$$\widetilde{\Theta}_{in}(x) = \chi(\delta(x)) \text{ and (for each } \nu) \; \widetilde{\Theta}_\nu(x) = (1 - \widetilde{\Theta}_{in}(x)) \cdot \Theta_\nu(x) \text{ for } x \in \mathbb{R}^n. \tag{17.2}$$

Thus

$$\widetilde{\Theta}_{in}, \widetilde{\Theta}_\nu \in C^\infty(\mathbb{R}^n), \qquad \widetilde{\Theta}_{in} \ge 0 \text{ and } \widetilde{\Theta}_\nu \ge 0 \text{ on } \mathbb{R}^n; \tag{17.3}$$

$$\widetilde{\Theta}_{in}(x) = 1 \text{ for } \delta(x) \le \eta; \tag{17.4}$$

$$\operatorname{supp} \widetilde{\Theta}_{in} \subset \{x \in \mathbb{R}^n : \delta(x) \le 2\eta\}; \tag{17.5}$$

$$\operatorname{supp} \widetilde{\Theta}_\nu \subset Q_\nu^* \text{ for each } \nu; \tag{17.6}$$

and

$$\widetilde{\Theta}_{in} + \sum_\nu \widetilde{\Theta}_\nu = 1 \text{ everywhere on } \mathbb{R}^n. \tag{17.7}$$

Note that (15.2) and (17.6) yield the following

$$\text{any given } x \in \mathbb{R}^n \text{ belongs to } \operatorname{supp} \widetilde{\Theta}_\nu \text{ for at most } C \text{ distinct } \nu. \tag{17.8}$$

Near Extensions and Alignment of Data in $\mathbb{R}^n$: Whitney extensions of near isometries, shortest paths, equidistribution, clustering and non-rigid alignment of data in Euclidean space,
First Edition. Steven B. Damelin.

In view of (17.5), we have

$$\operatorname{supp} \widetilde{\Theta}_{in} \subset U, \tag{17.9}$$

if η is small enough. This tells us in particular that $\widetilde{\Theta}_{in}(x) \cdot \phi(x)$ is a well-defined function from $\mathbb{R}^n$ to $\mathbb{R}^n$.

We establish the basic estimates for the gradients of $\widetilde{\Theta}_{in}$, $\widetilde{\Theta}_{\nu}$. By (17.4), (17.5) we have $\nabla\widetilde{\Theta}_{in}(x) = 0$ unless $\eta < \delta(x) < 2\eta$. For $\eta < \delta(x) < 2\eta$, we have

$$\left|\nabla\widetilde{\Theta}_{in}(x)\right| = \left|\chi'(\delta(x))\right| \cdot \left|\nabla\delta(x)\right| \leq C\eta^{-1}$$

by (17.1) and (15.9). Therefore,

$$\left|\nabla\widetilde{\Theta}_{in}(x)\right| \leq C(\delta(x))^{-1} \text{ for all } x \in \mathbb{R}^n \backslash E. \tag{17.10}$$

We turn our attention to $\nabla\widetilde{\Theta}_{\nu}(x)$. Recall that $0 \leq \Theta_{\nu}(x) \leq 1$ and $0 \leq \widetilde{\Theta}_{in}(x) \leq 1$ for all $x \in \mathbb{R}^n$. Moreover, (15.1), (15.4), (15.5) and (15.7) together yield

$$\left|\nabla\Theta_{\nu}(x)\right| \leq C\left(\delta(x)\right)^{-1} \text{ for all } x \in \mathbb{R}^n \backslash E \text{ and for all } \nu.$$

The above remarks (including (17.10)), together with the definition (17.2) of $\widetilde{\Theta}_{\nu}$, tell us that

$$\left|\nabla\widetilde{\Theta}_{\nu}(x)\right| \leq C(\delta(x))^{-1} \text{ for } x \in \mathbb{R}^n \backslash E, \text{ each } \nu. \tag{17.11}$$

18

Extensions of Smooth Small Distortions: Function Extension

We now define

$$\Phi(x) = \widetilde{\Theta}_{in}(x) \cdot \phi(x) + \sum_{\nu} \widetilde{\Theta}_{\nu}(x) \cdot A_{\nu}(x) \text{ for all } x \in \mathbb{R}^n. \tag{18.1}$$

This makes sense, thanks to (17.8) and (17.9). Moreover, $\Phi : \mathbb{R}^n \to \mathbb{R}^n$ is a C^1-function. We will prove that Φ satisfies all the conditions of Theorem 13.34.

First of all, for $\delta(x) < \eta$, (17.3), (17.4), (17.7) give $\widetilde{\Theta}_{in}(x) = 1$ and all $\widetilde{\Theta}_{\nu}(x) = 0$; hence (18.1) gives $\Phi(x) = \phi(x)$. Thus, Φ satisfies Theorem 13.34.

Next suppose $d(x) \geq c_0$. Then $\delta(x) > c > 2\eta$ if η is small enough; hence $\widetilde{\Theta}_{in}(x) = 0$ and $\widetilde{\Theta}_{\nu}(x) = \Theta_{\nu}(x)$ for each ν. (See (17.3) and (17.5).) Also, (15.10) shows that $\beta_{\nu} > c_3$ for all ν such that $x \in \operatorname{supp} \Theta_{\nu}$. For such ν, we have defined $A_{\nu} = A_{\infty}$; see (16.3). Hence, in this case,

$$\Phi(x) = \sum_{\nu} \Theta_{\nu}(x) \cdot A_{\infty}(x) = A_{\infty}(x),$$

thanks to (15.6). Thus, Φ satisfies Theorem 13.34.

Next, suppose $\phi \in C^m(U)$ for some given $m \geq 1$. Then since $\widetilde{\Theta}_{in}$ and each $\widetilde{\Theta}_{\nu}$ belong to $C^{\infty}(\mathbb{R}^n)$, we learn from (17.8), (17.9) and (18.1) that $\Phi : \mathbb{R}^n \to \mathbb{R}^n$ is a C^m function.

Similarly, if $\phi \in C^{\infty}(U)$, then $\Phi : \mathbb{R}^n \to \mathbb{R}^n$ is a C^{∞} function. Thus, Φ satisfies Theorem 13.34.

It remains to show that Φ satisfies Theorem 13.34. To establish these assertions, we first control $\nabla\Phi$.

Lemma 18.1. *For all $x \in \mathbb{R}^n$ such that $\delta(x) \leq 2\eta$, we have*

$$\left|\nabla\Phi(x) - \nabla\phi(x)\right| \leq C\varepsilon.$$

Near Extensions and Alignment of Data in $\mathbb{R}^n$: Whitney extensions of near isometries, shortest paths, equidistribution, clustering and non-rigid alignment of data in Euclidean space,
First Edition. Steven B. Damelin.

Proof. We may assume $\delta(x) \geq \eta$, since otherwise we have $|\nabla\Phi(x) - \nabla\phi(x)| = 0$. For $\delta(\underline{x}) \leq 3\eta$, we have $\underline{x} \in U$, and (18.1) gives

$$\Phi(\underline{x}) - \phi(\underline{x}) = \sum_{\nu} \widetilde{\Theta}_\nu(\underline{x})[A_\nu(\underline{x}) - \phi(\underline{x})], \tag{18.2}$$

since $\phi(\underline{x}) = \widetilde{\Theta}_{in}(\underline{x})\phi(\underline{x}) + \sum_\nu \widetilde{\Theta}_\nu(\underline{x})\phi(\underline{x})$. If $\delta(x) \leq 2\eta$, then (18.2) holds on a neighborhood of x; hence

$$\nabla\Phi(x) - \nabla\phi(x) = \sum_{\nu} \nabla\widetilde{\Theta}_\nu(x) \cdot [A_\nu(x) - \phi(x)] + \sum_{\nu} \widetilde{\Theta}_\nu(x) \cdot [\nabla A_\nu - \nabla\phi(x)]. \tag{18.3}$$

There are at most C non-zero terms on the right in (18.3), thanks to (17.8). Moreover, if η is small enough, then Lemma 16.2 and (17.6) show that $|A_\nu(x) - \phi(x)| \leq C\varepsilon\delta(x)$ and $|\nabla A_\nu - \nabla\phi(x)| \leq C\varepsilon$ whenever $\text{supp}\,\widetilde{\Theta}_\nu \ni x$. Also, for each ν, we have $0 \leq \widetilde{\Theta}_\nu(x) \leq 1$ by (17.3) and (17.7); and $|\nabla\widetilde{\Theta}_\nu(x)| \leq C \cdot (\delta(x))^{-1}$, by (17.11). Putting these estimates into (18.3), we obtain the conclusion of Lemma 18.1. ■

Lemma 18.2. *Let $x \in Q^*_\mu$, and suppose $\delta(x) > 2\eta$. Then*

$$\left|\nabla\Phi(x) - \nabla A_\mu\right| \leq C\varepsilon.$$

Proof. Since $\delta(x) > 2\eta$, we have $\widetilde{\Theta}_{in}(x) = 0$, $\nabla\widetilde{\Theta}_{in} = 0$, and $\widetilde{\Theta}_\nu(x) = \Theta_\nu(x)$, $\nabla\widetilde{\Theta}_\nu(x) = \nabla\Theta_\nu(x)$ for all ν; see (17.5) and (17.3). Hence, (18.1) yields

$$\nabla\Phi(x) = \sum_{\nu} \nabla\Theta_\nu(x)A_\nu(x) + \sum_{\nu} \Theta_\nu(x)\nabla A_\nu.$$

Since also

$$\nabla A_\mu = \sum_{\nu} \nabla\Theta_\nu(x)A_\mu(x) + \sum_{\nu} \Theta_\nu(x)\nabla A_\mu,$$

(as $\sum_\nu \nabla\Theta_\nu(x) = 0$, $\sum_\nu \Theta_\nu(x) = 1$), we have

$$\nabla\Phi(x) - \nabla A_\mu = \sum_{\nu} \nabla\Theta_\nu(x) \cdot [A_\nu(x) - A_\mu(x)] + \sum_{\nu} \Theta_\nu(x) \cdot [\nabla A_\nu - \nabla A_\mu]. \tag{18.4}$$

There are at most C non-zero terms on the right in (18.4), thanks to (17.8). By (15.1), (15.4), (15.5) and (15.7), we have $|\nabla\Theta_\nu(x)| \leq C(\delta(x))^{-1}$; and (15.3), (15.6) yield $0 \leq \Theta_\nu(x) \leq 1$. Moreover, whenever $Q^*_\nu \ni x$, Lemma 16.1 gives $|A_\nu(x) - A_\mu(x)| \leq C\varepsilon\delta(x)$, and $|\nabla A_\mu - \nabla A_\nu| \leq C\varepsilon$. When $Q^*_\nu \ni x$, we have $\Theta_\nu(x) = 0$ and $\nabla\Theta_\nu(x) = 0$, by (15.4). Using the above remarks to estimate the right-hand side of (18.4), we obtain the conclusion of Lemma 18.2. ■

Using Lemma 18.1 and 18.2, we can show that

$$(1 - C\varepsilon)I \leq \left(\nabla\Phi(x)\right)^{+}\left(\nabla\Phi(x)\right) \leq (1 + C\varepsilon)I \text{ for all } x \in \mathbb{R}^n. \tag{18.5}$$

Indeed, if $\delta(x) \leq 2\eta$, then (18.5) follows from Lemma 14.3 and (18.1). If instead $\delta(x) > 2\eta$, then $x \in \mathbb{R}^n \backslash E$, hence $x \in Q_\mu$ for some μ. Estimate (18.5) then follows from Lemma 18.2, since $(\nabla A_\mu)^+(\nabla A_\mu) = I$ for the Euclidean motion A_μ. Thus, (18.5) holds in all cases.

From (18.5), together with Lemma 14.5, we see that

$$\begin{aligned} &\Phi : \mathbb{R}^n \to \mathbb{R}^n \text{ is one-to-one and onto, hence} \\ &\Phi^{-1} : \mathbb{R}^n \to \mathbb{R}^n \text{ is a } C^1(\mathbb{R}^n) \text{ distorted-diffeomorphism} \end{aligned} \tag{18.6}$$

We use (18.5) and (18.6) as follows. Let $x, y \in \mathbb{R}^n$. Then $|x - y|$ is the minimum of length(Γ) over all C^1 curves Γ joining x to y. Also, by (18.6), $|\Phi(x) - \Phi(y)|$ is the infimum of length$(\Phi(\Gamma))$ over all C^1 curves Γ joining x to y. For each Γ, (18.5) yields

$$(1 - C\varepsilon)\,\text{length}\,(\Gamma) \leq \text{length}\,\left(\Phi(\Gamma)\right) \leq (1 + C\varepsilon)\,\text{length}\,(\Gamma).$$

Taking the minimum over all Γ, we conclude that Φ satisfies all the conditions of Theorem 13.34 completing the proof. ■

19

Equidistribution: Extremal Newtonian-like Configurations, Group Invariant Discrepancy, Finite Fields, Combinatorial Designs, Linear Independent Vectors, Matroids and the Maximum Distance Separable Conjecture

We use the notation μ and ν for measures.

The problem of "distributing well" a large number of points on certain n-dimensional compact sets embedded in $\mathbb{R}^{n+1}$ (see Figures 19.1–19.6) is an interesting problem with numerous wide applications in diverse areas, for example, harmonic analysis, approximation theory, zeroes of extremal polynomials in all kinds of settings, singular operators, for example, Hilbert transforms, random matrix theory, crystal and molecule structure, electrostatics, special functions, Newtonian energy, extensions, alignment, data science, number theory, manifold learning, clustering, shortest paths, codes and discrepancy, computer vision, signal processing, biology, neuroscience, networks, clustering, optimal transport, and many others. We say the points equidistribute over the set or cover it.

We now discuss equidistribution and covers of certain n-dimensional compact sets embedded in $\mathbb{R}^{n+1}$ via extremal Newtonian-like configurations, group invariant discrepancy, finite fields, combinatorial designs, linear independent vectors, matroids, and the maximum distance separable (MDS) conjecture. See our work [7, 11, 28, 32, 43, 46, 48–51, 54, 55, 109]

19.1 s-extremal Configurations and Newtonian s-energy

Consider the circle S^1 and a configuration of $k \geq 2$ points on S^1 being the vertices of the regular k-gon. How to make sense of the following generalization?

A configuration with $k \geq 2$ points is "distributed well" on a compact set, an element of a certain class of n-dimensional compact sets embedded in $\mathbb{R}^{n+1}$?

Near Extensions and Alignment of Data in $\mathbb{R}^n$: Whitney extensions of near isometries, shortest paths, equidistribution, clustering and non-rigid alignment of data in Euclidean space,
First Edition. Steven B. Damelin.

Let us be given an n-dimensional compact set X embedded in $\mathbb{R}^{n+1}$ and a $k \geq 2$ configuration $\omega_k = \{x_1, ..., x_k\}$ on X. The discrete Newtonian s-energy associated with ω_k is given by

$$E_s(X, \omega_k) := \sum_{1 \leq i < j \leq k} |x_i - x_j|^{-s}.$$

when $s > 0$ and

$$E_s(X, \omega_k) := \sum_{1 \leq i < j \leq k} \log |x_i - x_j|^{-1}$$

when $s = 0$. Let $\omega_s^*(X, k) := \{x_1^*, ..., x_k^*\} \subset X$ be a configuration for which $E_s(X, \omega_k)$ attains its minimal value, that is,

$$\mathcal{E}_s(X, k) := \min_{\omega_k \subset X} E_s(X, \omega_k) = E_s(X, \omega_s^*(X, k)).$$

We call such minimal configurations s-extremal configurations.

It is well-known that in general s-extremal configurations are not always unique. For example, in the case of the n-dimensional unit sphere S^n embedded in $\mathbb{R}^{n+1}$ they are invariant under rotations.

In this section, we are interested in how s-extremal configurations distribute (for large enough k) on the interval $[-1, 1]$, on the n-dimensional sphere S^n embedded in $\mathbb{R}^{n+1}$ and on the n-dimensional torus embedded in $\mathbb{R}^{n+1}$. We are also interested in separation radius and mesh norm estimates of $k > 2$ s-extremal configurations on a class of n-dimensional compact sets embedded in $\mathbb{R}^{n+1}$ with positive Hausdorff measure. We will define this class of compact sets, and terms separation radius and mesh norm. Asymptotics of energies $\mathcal{E}_s(X, k)$ for large enough k on certain n-dimensional compact sets embedded in $\mathbb{R}^{n+1}$ is an interesting topic of study but we do not address this topic in this book.

19.2 [−1,1]

19.2.1 Critical Transition

Let us look at the interval $[-1, 1]$. Here, identifying as usual $[-1, 1]$ as the circle S^1, $[-1, 1]$ has dimension $n = 1$ and is embedded in $\mathbb{R}^2$. $[-1, 1]$ has Hausdorff measure 1.

(1) In the limiting s cases, i.e., $s = 0$ (logarithmic interactions) and $s = \infty$ (best-packing problem), s-extremal configurations are Fekete points and equally spaced points, respectively.

(2) Fekete points are distributed on $[-1,1]$ for large enough k according to the arcsine measure, which has the density $\mu_0'(x) := (1/\pi)(1-x^2)^{-1/2}$.
(3) Equally spaced points, $-1+2(i-1)/(k-1)$, $i=1,...,k$, have the arclength distribution for large enough k.
(4) $s=1$ is a critical transition in the sense that s-extremal configurations are distributed on $[-1,1]$ for large enough k differently for $s <,1$ and $s \geq 1$. Indeed, for $s<1$, the limiting distribution of s-extremal configurations for large enough k is an arcsin type density $\mu_s(x) = \frac{\Gamma(1+s/2)}{\sqrt{\pi}\Gamma((1+s/2)/2)}(1-x^2)^{(s-1)/2}$ where Γ is as usual the Gamma function.

19.2.2 Distribution of s-extremal Configurations

The dependence of the distribution of s-extremal configurations over $[-1,1]$ for large enough k and asymptotics for minimal discrete s-energy can be easily explained from a potential theory point of view as follows.

For a probability Borel measure ν on $[-1,1]$, its s-energy integral is defined to be

$$I_s([-1,1],\nu) := \iint\limits_{[-1,1]^2} |x-y|^{-s} d\nu(x)d\nu(y)$$

(which can be finite or infinite).

Let now, for a $k \geq 2$ configuration $\omega_k = \{x_1,...,x_k\}$ on $[-1,1]$,

$$\nu^{\omega_k} := \frac{1}{k}\sum_{i=1}^{k} \delta_{x_i}$$

denote the normalized counting measure of ω_k so that $\nu^{\omega_k}([-1,1]) = 1$. Then the discrete Newtonian s-energy, associated to ω_k can be written as

$$E_s([-1,1],\omega_k) = \frac{1}{2k^2}\iint\limits_{x\neq y} |x-y|^{-s} d\nu^{\omega_k}(x)d\nu^{\omega_k}(y)$$

where the integral represents a discrete analog of the s-energy integral for the point-mass measure ν^{ω_k}.

We observe the following: if $s<1$, then the energy integral is minimized uniquely by an arcsine-type measure ν_s^*, with density $\mu_s'(x)$ with respect to the Lebesgue measure. On the other hand, the normalized counting measure $\nu_{s,k}^*$ of an s-extreme configuration minimizes the discrete energy integral over all configurations ω_k on $[-1,1]$. Thus, one can reasonably expect that for k large enough $\nu_{s,k}^*$ is close to ν_s^* for certain s.

Indeed, we find that for $s \geq 1$, the energy integral diverges for every measure ν. Concerning the distribution of s-extremal points over $[-1, 1]$ for very large k, the interactions are now strong enough to force the points to stay away from each other as far as possible. Of course, depending on s, "far" neighbors still incorporate some energy in $\mathcal{E}_s([-1, 1], k)$, but the closest neighbors are dominating. So, s-extremal points distribute themselves over $[-1, 1]$ in an equally spaced manner for large enough k.

See below where we study this idea on the sphere S^n as a discrepancy of measures.

19.2.3 Equally Spaced Points for Interpolation

As an indication of how "equally spaced" points can be far from ideal in many approximation frameworks, it is well-known that equally spaced points for example on the interval $[-1, 1]$ can be disastrous for certain one variable approximation processes such as interpolation. Indeed, a classical result of Runge for example shows this.

19.3 The n-dimensional Sphere, S^n Embedded in $\mathbb{R}^{n+1}$

19.3.1 Critical Transition

S^n has positive Hausdorff measure n. It turns out that for any s the limiting distribution of s-extremal configurations on S^n for large enough k is given by the normalized area measure on S^n. This is due to rotation invariance.

Consider the sphere S^2 embedded in $\mathbb{R}^3$. The s-extremal configurations presented are close to global minimum. In the table below, ρ denotes mesh norm (fill distance), $2\hat{\rho}$ denotes separation angle which is twice the separation (packing) radius and a denotes mesh ratio which is ρ/ρ'. We recall the concepts of separation radius and mesh norm.

The s-extremal configurations for $s = 1, 2, 3, 4$ ($s < n$, $s = n$, $s > n$, $n = 2$) given in plots 1–4 are for 400 points respectively. Given the symmetry, the points are similar for all values of s considered

s,	ρ,	$2\hat{\rho}$,	a
1,	0.113 607,	0.175 721,	1.2930
2,	0.127 095,	0.173 361,	1.4662
3,	0.128 631,	0.173 474,	1.4830
4,	0.134 631,	0.172 859,	1.5577

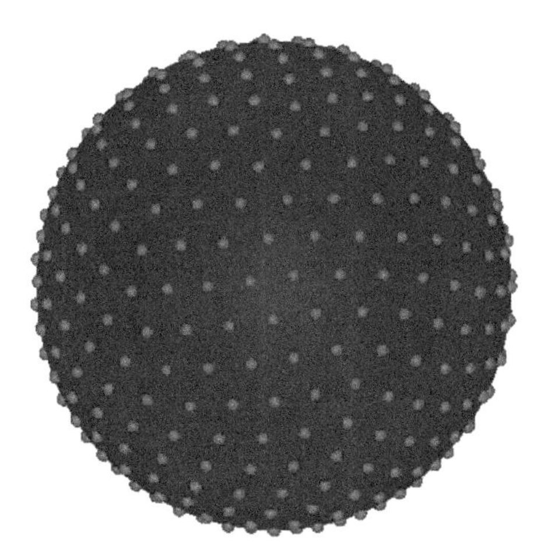

Figure 19.1 S^2, $s = 1$.

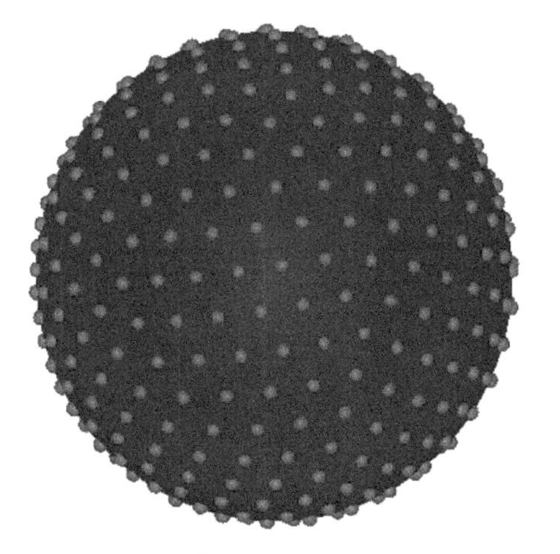

Figure 19.2 S^2, $s = 2$.

19.4 Torus

Consider a torus embedded in $\mathbb{R}^3$ with inner radius 1 and outer radius 3. In this case, we no longer have symmetry and we have three transition cases. The s-extremal configurations for $s = 1, 2, 3$ ($s < n$, $s = n$, $s > n$, $n = 2$) respectively are not similar. Again, we have 400 points.

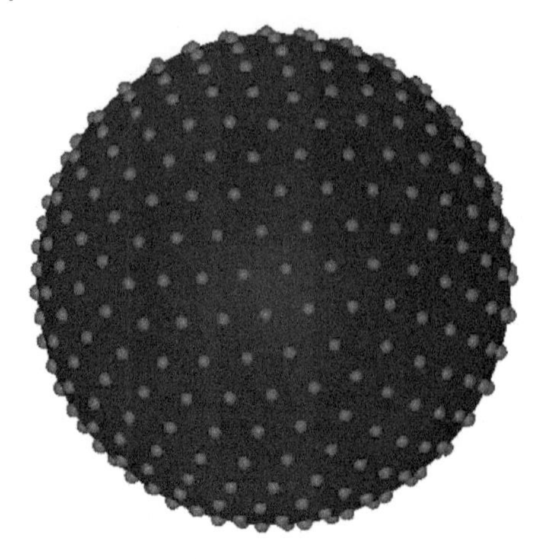

Figure 19.3 S^2, $s = 3$.

Figure 19.4 Torus, $s = 1$.

19.5 Separation Radius and Mesh Norm for s-extremal Configurations

We are now interested in the separation radius (packing radius) and mesh norm (fill-distance) of $k \geq 2$ s-extremal configurations on n-dimensional compact sets embedded in $\mathbb{R}^{n+1}$ with positive n-dimensional Hausdorff measure and which are a finite union of bi-Lipschitz images of compact sets in $\mathbb{R}^n$. We call any element of this class of compact sets Y^n where as usual the notation Y^n may denote the same or different set in different occurrences.

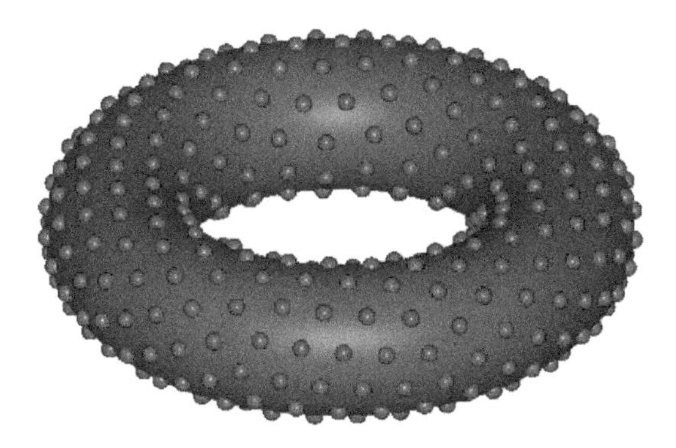

Figure 19.5 Torus, $s = 2$.

Figure 19.6 Torus, $s = 3$.

Examples of these compact sets Y^n are n-dimensional spheres, n-dimensional tori, and n-dimensional ellipsoids embedded in $\mathbb{R}^{n+1}$. Other examples are n-dimensional balls, n-dimensional cubes, and parallelepipeds, n-dimensional Cantor sets having positive n-dimensional Hausdorff measure and quasi-smooth (chord-arc) curves in $\mathbb{R}^{n+1}$.

For $j = 1, ..., k$ and a configuration $\omega_k = \{x_1, ..., x_k\}$ of distinct points on a given set Y^n, we let

$$\hat{\rho}_j(\omega_k) := \min_{i \neq j} \{|x_i - x_j|\} \tag{19.1}$$

and define

$$\hat{\rho}(\omega_k) := \min_{1 \leq j \leq k} \delta_j(\omega_k). \tag{19.2}$$

The quantity $\hat{\rho}(\omega_k)$ is called the *separation radius or packing radius* of the configuration ω_k and gives the minimal distance between points in the configuration ω_k on the given set Y^n.

We also define the *mesh norm or fill distance* of the given configuration ω_k on the given set Y^n denoted by $\rho(Y^n, \omega_k)$ to be the maximal radius of a cap on Y^n, which does not contain points from ω_k. It is defined as

$$\rho(Y^n, \omega_k) := \max_{y \in Y^n} \min_{x \in \omega_k} |y - x|. \tag{19.3}$$

Here are our results for $k \geq 2$ s-extremal configurations.

19.5.1 Separation Radius of $s > n$-extremal Configurations on a Set Y^n

The following hold.

Theorem 19.3. *Let a Y^n be given. For any $s > n$-extremal configuration $\omega_s^*(Y^n, k)$ on Y^n,*

$$\hat{\rho}_s^*(Y^n, k) := \hat{\rho}(\omega_s^*(Y^n, k)) \geq ck^{-1/n}. \tag{19.4}$$

19.5.2 Separation Radius of $s < n - 1$-extremal Configurations on S^n

Separation results for weak interactions $s < n$ are far more difficult to find in the literature for sets Y^n. A reason for such a lack of results is that this case requires delicate considerations based on the minimizing property of $\omega_s^*(Y^n, k)$ while strong interactions $(s > n)$ prevent points from being very close to each other without affecting the total energy.

Theorem 19.4.

(1) *For $n \geq 2$ and $s < n - 1$,*

$$\hat{\rho}_s^*(S^n, k) \geq ck^{-1/(s+1)}. \tag{19.5}$$

(2) *The estimate above can be improved for $n \geq 3$ and $s \leq n - 2$ in the following sense.*

$$\hat{\rho}_s^*(S^n, k) \geq ck^{-1/(s+2)}. \tag{19.6}$$

19.5.3 Mesh Norm of s-extremal Configurations on a Set Y^n

In this section, we will obtain for $s > n$ an upper bound of $O(k^{-1/n})$ for $\rho_s^*(Y^n, k) := \rho(Y^n, \omega_s^*)$ matching the lower bound given for $\hat{\rho}_s^*(Y^n, k)$ for a given Y^n. Let a set Y^n be given, let $x \in Y^n$ and let a radius $r > 0$ be given. Then we define a cap on the set Y^n with center x and radius r by $\text{cap}(x,r) := \{y \in Y^n : |y-x| < r\}$. We have seen the notion of a spherical cap earlier.

The problem in getting upper bounds for $\rho_s^*(Y^n, k)$ is the following observation. For any $s > 0$ and any $k \geq 2$ configuration ω_k on Y^n with

$$\lim_{k\to\infty} \frac{E_s(Y^n, \omega_k)}{\mathcal{E}_s(Y^n, k)} = 1 \tag{19.7}$$

it follows that

$$\lim_{k\to\infty} \rho(Y^n, \omega_k) = 0. \tag{19.8}$$

Thus an "optimistic" guess is that we require an additional assumption on Y^n to get a reasonable upper bound for $\rho_s^*(Y^n, k)$ and indeed a matching one of $O(k^{-1/n})$. This turns out to be case.

We have:

Theorem 19.5. *Let a Y^n be given. Then the following holds:*

(1) *Choose $x \in Y^n$ and a radius r. Then the n-dimensional Hausdorff measure of $\text{cap}(x,r)$ is $O(r^n)$ uniformly in x, r.*
(2) *Suppose now we have the matching lower bound in (1) in the sense of the following assumption: Choose $x \in Y^n$ and a radius r. Suppose there exists $c > 0$ depending on n but independent of x, r so that the n-dimensional Hausdorff measure of $\text{cap}(x,r)$ is bounded below by cr^n uniformly in x, r. Then the following matching bound for $\hat{\rho}_s^*(Y^n, k)$ holds.*

$$\rho(Y^n, \omega_s^*(Y^n, k)) = O(k^{-1/n}). \tag{19.9}$$

We refer the reader to the papers [28, 29, 33, 50, 51] for more details of the work in this section. For example, as already stated, we provide no analysis on the behavior of $E_s(Y^n, \omega_k)$ and $\mathcal{E}_s(Y^n, k)$ for different s, Y^n, ω_k which provides some very interesting results. Sarnak and his collaborators have studied local statistics of lattice points. An interesting question would be to see if any of their work can be applied to our work. For a configuration ω_k, $k \geq 1$ of points randomly and independently distributed by area measure on S^n, it is known that

$$\text{Expect}((\rho(S^n, \omega_k)) = O((\log k)^{-1/n})) \tag{19.10}$$

uniformly for large enough k. Here $\text{Expect}(\cdot)$ is statistical expectation.

19.6 Discrepancy of Measures, Group Invariance

We guess optimistically that for certain $s > 0$ and k large enough, the measures $\nu_{s,k}^*$ should be close to $\nu_s^* = \mu$ on the sphere S^n. Here μ is the normalized surface measure on the sphere S^n. We write $\omega_s^*(S^n,k) = \{x_1, ..., x_k\}$. We define

$$R(f, \omega_s^*(S^n,k), \mu) := \left| \int_{S^n} f(x)d\mu(x) - \frac{1}{k} \sum_{1 \le i \le k} f(x_i) \right| \tag{19.11}$$

where $f : S^n \to \mathbb{R}$ is continuous.

Indeed, we have as one of our results:

Theorem 19.6. *Let $\omega_s^*(S^n,k)=\{x_1,...,x_k\}$ an s-extremal configuration on the n-dimensional sphere S^n. Suppose that $s>n$, $f:S^n \to \mathbb{R}$ is continuous, $R(f,\omega_s^*(S^n,k),\mu)$ is defined as above and μ is the normalized surface measure on the sphere S^n. Then for k large enough,*

$$R(f, \omega_s^*(S^n,k), \mu) = O\left(\left(\frac{1}{\sqrt{\log k}}\right)\right). \tag{19.12}$$

The constant in the upper bound depends on f, s, n. The error $R(\cdot,\cdot)$ is called the discrepancy of the two measures: $\nu_{s,k}^*$ and $\nu_s^* = \mu$. Discrepancy theory is a large subject of study with many applications, for example, numerical analysis, approximation theory, number theory, probability, ergodic theory, and many others. Regarding our works above, see the papers [28, 32, 43, 46, 49–51].

In [46, 49], we analyzed the question of discrepancy and energy in a group invariance framework for a measurable subset X of $\mathbb{R}^n$ as the orbit of a compact, possibly non-Abelian group, G, acting as measurable transformations of X. The papers study discrepancy and energy spaces with transitive measurable group actions as examples:

(1) The unit sphere S^n as a subset of $\mathbb{R}^{n+1}$ as the orbit of any unit vector under the action of $SO(n+1)$.
(2) The flat torus $T^n = (S^1)^n$ as a subset of $(\mathbb{R}^2)^n$ which is the orbit of the point $((1,0),(1,0),...,(1,0))$ under rotation by a vector of angles $(\theta_1,\theta_2,...,\theta_n)$ which is the product group $(\mathbb{R}/2\pi\mathbb{Z})^n$.
(3) The non-flat 2-torus in $\mathbb{R}^3$ given by $\{(x,y,z)\}$ with

$$x = (r_1 + r_2\sin(\theta_2))\cos(\theta_1),\ y = (r_1 + r_2\sin(\theta_2))\sin(\theta_1)$$

$$z = r_2\cos(\theta_2),\ 0 \le \theta_1 < 2\pi, 0 \le \theta_2 < 2\pi$$

with $r_2 > r_1 > 0$ fixed. The group $(\mathbb{R}/2\pi\mathbb{Z})^2$ acts transitively via translation in the θ_1, θ_2 coordinates.
(4) An important variant of (3). The open unit n-cube $[0,1)^n$ admitting a transitive measurable action of the compact group $(\mathbb{R}/\mathbb{Z})^n$ given by translation mod 1.

Discrepancy on compact, homogeneous manifolds embedded in $\mathbb{R}^{n+1}$, via energy functionals associated with a class of group-invariant kernels which are generalizations of so-called zonal kernels on spheres or radial kernels are also studied in depth. For example, certain weighted Newtonian kernels on spheres and certain projective spaces. We refer the reader to the papers [46, 49] for the exact details.

19.7 Finite Field Algorithm

In this section, we provide a description of some of our work in [7]. Here, we show from [7] a finite field algorithm to generate "well distributed" point sets on the n-dimensional sphere S^n. Well distributed for us is measured via discrepancy.

We proceed as follows.

For an odd prime $\hat{p}$, let $F_{\hat{p}}$ denote the finite field of integers modulo $\hat{p}$. We consider the quadratic form

$$x_1^2 + ... + x_{n+1}^2 = 1. \tag{19.13}$$

over $F_{\hat{p}}$ for a given prime $\hat{p}$.

Step 1 Let $k(n, \hat{p})$ denote the number of solutions of this quadratic form for such $\hat{p}$. The number of solutions is known to be given as

$$k(n, \hat{p}) = \begin{cases} \hat{p}^n - \hat{p}^{(n-1)/2}\eta((-1)^{(n+1)/2}) & \text{if } n \text{ is odd} \\ \hat{p}^n + \hat{p}^{n/2}\eta((-1)^{n/2}) & \text{if } n \text{ is even.} \end{cases}$$

Here η is the quadratic character defined on $F_{\hat{p}}$ by $\eta(0) = 0$, $\eta(a) = 1$ if a is a square in $F_{\hat{p}}$, and $\eta(a) = -1$ if a is a non-square in $F_{\hat{p}}$.

Thus the number of solutions to the form above for a given prime $\hat{p}$ is given by the number $k(n, \hat{p})$.

Step 2 We now scale and center around the origin. Given a prime $\hat{p}$, we let $X(n, \hat{p})$ be a solution vector of the form above. We write it as

$$X(n, \hat{p}) = (x_1, ..., x_{n+1}),\ x_i \in F_{\hat{p}},\ 1 \leq i \leq n + 1,$$

with coordinates x_i, $1 \leq i \leq n+1$. We may assume without loss of generality that all coordinates x_i of the vector $X(n, \hat{p})$ are scaled so that they are centered around the origin and are contained in the set

$$\{-(\hat{p} - 1)/2, ..., (\hat{p} - 1)/2\}.$$

More precisely given a coordinate x_i of the vector $X(n, \hat{p})$ for some $1 \leq i \leq n+1$, define

$$x_i' = \begin{cases} x_i, & x_i \in \{0, ..., (\hat{p} - 1)/2)\} \\ x_i - \hat{p}, & x_i \in \{(\hat{p} + 1)/2, ..., \hat{p} - 1\}. \end{cases}$$

Then $x'_i \in \{-(\hat{p}-1)/2, ..., (\hat{p}-1)/2\}$ and the scaled vector

$$X'(n,\hat{p}) = (x'_1, ..., x'_{D+1}),\ 1 \le i \le n+1$$

solves the form above, if and only if the vector $X(n,\hat{p})$ solves the form above.

Step 3 We now simply normalize the coordinates of the vector $X'(n,\hat{p})$ so that the vector we end up with is on S^n. For convenience we denote this vector by $X(n,\hat{p})$. We have $k(n,\hat{p})$ such vectors as $\hat{p}$ varies.

Use of the finite field $F_{\hat{p}}$ provides a method to increase the number $k(n,\hat{p})$ of vectors $X(n,\hat{p})$ as $\hat{p}$ increases. For increasing values of $\hat{p}$, we obtain an increasing number $k(n,\hat{p}) = O(\hat{p}^n)$ of vectors scattered on S^n. In particular, as $\hat{p} \to \infty$ through all odd primes, it is clear that $k(n,\hat{p}) \to \infty$.

19.7.1 Examples

We begin by seeing one straightforward way to produce a full set of vectors $X(n,\hat{p})$ for a given prime $\hat{p}$. Fix a prime $\hat{p}$ and consider its corresponding vectors $X''(n,\hat{p})$. Now take ± 1 times all coordinates of the vectors $X''(n,\hat{p})$ permuting each coordinate in all possible ways. This procedure produces the required full set of $k(n,\hat{p})$ vectors $X(n,\hat{p})$ for the given prime $\hat{p}$. The table below produces vectors $X''(n,\hat{p})$ and the number $k''(n,\hat{p})$ of the full set of vectors $X(n,\hat{p})$ for the three odd primes 3, 5, 7 and for dimensions $n = 1, 2$

D	$\hat{p}$	$k(n,\hat{p})$	$X''(n,\hat{p})$
1	3	4	$\{(1,0)\}$
1	5	4	$\{(1,0)\}$
1	7	8	$\{(1,0), \frac{1}{\sqrt{2}}(1,1)\}$
2	3	6	$\{(1,0,0)\}$
2	5	30	$\{(1,0,0), \frac{1}{\sqrt{2}}(2,1,1)\}$
2	7	42	$\{(1,0,0), \frac{1}{\sqrt{2}}(1,1,0), \frac{1}{\sqrt{22}}(3,3,2)\}$

Observe that for $\hat{p} = 3, 5, 7$ and $n = 1$, the finite field algorithm gives the optimal point set of vectors $X(1,\hat{p})$ in the sense that the $k(1,\hat{p})$ vectors $X(1,\hat{p})$ are precisely the vertices of the regular $k(1,\hat{p})$-gon on S^1. The $k(1,\hat{p})$ vectors $X(1,\hat{p})$ for $\hat{p} > 7$ do not always exhibit this feature.

19.7.2 Spherical $\hat{t}$-designs

Let $\hat{t}$ be a positive integer. A finite set of points X on S^n is a *spherical $\hat{t}$-design* or a *spherical design of strength $\hat{t}$*, if for every polynomial f of total degree $\hat{t}$ or less, the average value of f over the whole sphere S^n is equal to the arithmetic average of its values on X. If this only holds for homogeneous polynomials of degree $\hat{t}$, then X is called a *spherical design of index* $\hat{t}$.

The following holds:

Theorem 19.7. *For every odd positive integer $\hat{t}$ and odd prime $\hat{p}$, the set of $k(n, \hat{p})$ vectors $X(n, \hat{p})$ obtained from the finite field algorithm is a spherical design of index $\hat{t}$. Furthermore, each vector $X(n, \hat{p})$ is a spherical 3-design.*

We see then that in the sense of discrepancy, the set of $k(n, \hat{p})$ vectors $X(n, \hat{p})$ is well distributed over S^n.

19.7.3 Extension to Finite Fields of Odd Prime Powers

We now study a finite field algorithm to finite fields of odd prime powers. We proceed as follows.

Solve the same quadratic form over a finite field $F_{\hat{q}}$, where $\hat{q} = (\hat{p})^{\hat{e}}$ and where $\hat{e}$ is an odd prime.

The field $F_{\hat{q}}$ is a $\hat{e}$-dimensional vector space over the field $F_{\hat{p}}$. Thus, let $e_1, ..., e_{\hat{e}}$ be a basis of $F_{\hat{q}}$ over $F_{\hat{p}}$. If $x \in F_{\hat{q}}$, then x can be uniquely written as $x = (\hat{e'})_1 e_1 + \cdots + (\hat{e'})_{\hat{e}} e_{\hat{e}}$, where each $\hat{e'}_i \in F_p$, $1 \leq i \leq \hat{e}$. Moreover, we may assume as before that each $(\hat{e'})_i$, $1 \leq i \leq \hat{e}$ satisfies $-(p-1)/2 \leq \hat{e'}_i \leq (p-1)/2$ for $1 \leq i \leq \hat{e}$.

If the vector $(x_1, ..., x_{n+1})$ is a solution to the quadratic form over $F_{\hat{q}}$, then each coordinate x_i, $1 \leq i \leq n+1$ of this vector can be associated to an integer $(\hat{e''})_i = (\hat{e'})_{1,i} + (\hat{e'})_{2,i}(\hat{p}) + \cdots + (\hat{e'})_{\hat{e},i}(\hat{p})^{\hat{e}-1}$ for $1 \leq i \leq n+1$. It is an easy exercise to check that indeed $-(p^{\hat{e}} - 1)/2 \leq (\hat{e''})_i \leq (p^{\hat{e}} - 1)/2$ for each $1 \leq i \leq n+1$.

We then place the vector $((\hat{e''})_1, ..., (\hat{e''})_{n+1})$ to the surface of the unit sphere S^n by normalizing. For increasing values of $\hat{e}$, we obtain an increasing number $k_{\hat{e}}$ points scattered on the surface of S^n, so that as $\hat{e} \to \infty$, it is clear that $k_{\hat{e}} \to \infty$. We note that when $\hat{e} = 1$, our new construction reduces to our original construction.

19.8 Combinatorial Designs, Linearly Independent Vectors, MDS Conjecture

Our work in [7, 48, 54, 55, 109] relates to combinatorial designs, linearly independent vectors, MDS conjecture and matroids. All of the combinatorial designs, linearly independent vectors and matroid constructions lead to points with good discrepancy for various n-dimensional compact sets embedded in $\mathbb{R}^{n+1}$, the typical example S^n, the sphere of unit radius. The examples studied and related to combinatorial designs are orthogonal hypercubes, net designs, orthogonal arrays and linear codes. See our paper [55] and the classic Niederreiter et al. [89, 90].

An important idea for the work described on combinatorial designs relates to a problem of computing the cardinality of certain sets of linear independent vectors and then using these vectors scaled appropriately to construct these designs and a priori points of good discrepancy and covering. Our work on studying the cardinality of sets of linear independent vectors is that of [54, 55, 109] and the connections to combinatorial designs as of the above is in [55].

Here, now, we center on our work of [54, 55] related to linear independence. We state the main results with no proofs.

The quantity $\mathrm{Ind}_q(n,k)$.

Let k and n be integers with $2 \le k \le n$ and let q be a prime power. A set of vectors is k-independent if all its subsets with no more than k elements are linearly independent. We denote by $\mathrm{Ind}_q(n,k)$, the maximal possible cardinality of a k-independent set of vectors in the n-dimensional vector space F_q^n of order q. For example, in the case when we work with the binary case $q = 2$, F_2^n is the set of all vectors of length n consisting of 1's and 0's (i.e., n entries). Estimates on the size of $\mathrm{Ind}_q(n,k)$ provide information on the existence of the different combinatorial designs above, see [55]. It follows quite easily that the following hold for $\mathrm{Ind}_q(n,k)$.

$$q^n - 1 = \mathrm{Ind}_q(n,1) \ge \mathrm{Ind}_q(n,2) \ge \dots \ge \mathrm{Ind}_q(n,n) \ge n+1$$

and

$$\mathrm{Ind}_2(n,1) = \mathrm{Ind}_2(n,2).$$

19.8.1 The Case $q = 2$

Moving forward there are two extreme cases: $k \le 3$ and $k \ge 2n/3$. The main theorem in [55] is then the following:

Theorem 19.8. *The following hold:*

(a) $Ind_2(n,1) = Ind_2(n,2) = 2^n - 1$.
(b) $Ind_2(n,3) = 2^{n-1}$.
(c) $Ind_2(n,n-m) = n+1,\ n \ge 3m+2,\ m \ge 0$.
(d) $Ind_2(n,n-m) = n+2,\ n = 3m+i, i = 0,1, m \ge 2$.

19.8.2 The General Case

The next theorem generalizes Theorem 19.8 to F_q^n. This is the main result in [54].

Theorem 19.9. *The following hold:*

(a) $Ind_q(n,2) = \frac{q^n-1}{q-1}$.
(b) $Ind_q(n,k) = n+1$ *iff* $\frac{q}{q+1}(n+1) \le k$

Note that Theorem 19.9 part (a) with $q = 2$ is Theorem 19.8 part (a). Also note that Theorem 19.9 part (b) says that Theorem 19.8 part (c) is an iff statement.

19.8.3 The Maximum Distance Separable Conjecture

The following work appears in [48], details can be found there.

Let $k \geq 2$ be a positive integer, $n > 1$, a positive integer, q be a power of a prime p, and r a positive integer. A $k \times n$ maximum distance separable ($k \times n$ MDS) code M is a $k \times n$ matrix with entries in $\mathbb{F}_q$ such that every set of k columns of M is linearly independent. The MDS conjecture is a well-known problem in coding theory and algebraic geometry with important consequences for example to the study of arcs in finite projective spaces [12] and to coding theory [13, 14]. An excellent survey on the MDS conjecture can be found in [12].

The conjecture, first posed by Singleton in 1964 [108], gives a possible upper-bound on the size of a $k \times n$ MDS code. More precisely, the MDS conjecture says the following:

The maximum width, n of a $k \times n$ MDS code with entries in $\mathbb{F}_q$ is $q + 1$, unless q is even and $k \in \{3, q-1\}$, in which case the maximum width is $q + 2$. The conjecture is known to hold when q is a square and with $k \leq c\sqrt{pq}$ where the constant c depends on whether q is odd or even. When q not a square, it is known to be true for $k \leq c'\sqrt{pq}$ where c' depends on whether q is odd or even. It is also known to hold for all $k \times n$ MDS codes with alphabets of size at most eight. Our work is motivated by the following strongest result on the MDS conjecture which currently holds. [12].

Theorem 19.10. *Let k be an integer such that $2 \leq k \leq q = p^r$, where p is prime and r is a positive integer. The MDS conjecture is true whenever $k \leq 2p - 2$.*

For our contribution to this conjecture, we provide some needed notation.

(a) We denote by $\mathcal{P}_q$ the ring $\mathbb{F}_q[x]/(x^q - x)$ of polynomial functions over $\mathbb{F}_q$ of maximum degree q. $\mathcal{P}_q$ is a vector space over $\mathbb{F}_q$.

(b) Throughout, the perp space of a vector, $\boldsymbol{v}$ in an inner product space, $(\mathbf{V}, \cdot) := \mathbf{V}$ is the set given by $\boldsymbol{v}^\perp = \{\boldsymbol{w} \in \mathbf{V} | \boldsymbol{w} \cdot \boldsymbol{v} = 0\}$.

The *perp space* of a subspace, $\mathbf{U} \subseteq \mathbf{V}$, where $\mathbf{V}$ is an inner product space is the set given by $\mathbf{U}^\perp = \{\boldsymbol{w} \in \mathbf{V} | \forall \boldsymbol{u} \in \mathbf{U}, \boldsymbol{w} \cdot \boldsymbol{u} = 0\}$. ($\boldsymbol{w}$ and $\boldsymbol{u}$ are orthogonal).

(c) For every non-negative integer n, we define the subset $\mathcal{O}_n \subset \mathcal{P}_q$ as the set of polynomials in $\mathcal{P}_q$ that are either the zero polynomial, or have at most n distinct roots in $\mathbb{F}_q$. If $n \geq q$, then $\mathcal{O}_n = \mathcal{P}_q$.

(d) We denote by $\langle Y, Z \rangle$ the subspace of $\mathcal{P}_q$ generated by the elements of subspaces Y and Z.

(e) A *Reed–Solomon code* of dimension k is a $k \times q$ matrix with entries in $\mathbb{F}_q$ such that each column of the matrix is of the form $[1, a, a^2, ..., a^{k-2}, a^{k-1}]^\intercal$ for some $a \in \mathbb{F}_q$.

(f) An *extended Reed–Solomon code* is a Reed–Solomon code with the column $[0, 0, 0, ..., 0, 1]^\top$ appended.

Both Reed–Solomon and extended Reed–Solomon codes are $k \times n$ MDS codes, see [12] and it is known that for odd q, no column other than $[0, 0, 0, ..., 0, 1]^\top$ can be appended to a Reed–Solomon code to produce another $k \times n$ MDS code.

We have as our main result from [48]

Theorem 19.11. *Let k be an integer such that $2 \leq k \leq q = p^r$, where p is prime and r is a positive integer. Suppose that either*

- *q odd*
- *q even and $k \in \{3, q-1\}$.*

Consider the following statements (1–5) below:

(1) The MDS conjecture is true. That is, the maximum width, n, of a $k \times n$ MDS code with entries in $\mathbb{F}_q$ is $q+1$, unless q is even and $k \in \{3, q-1\}$, in which case the maximum width is $q+2$.
(2) Let M' be a $k \times (q+2)$ matrix. Then some linear combination of the rows of M' has at least k zero entries.
(3) Let M' be a $k \times (q+2)$ matrix such that the first two columns of M' are $[1, 0, ..., 0]^\top$ and $[0, 1, 0, ..., 0]^\top$. Then some linear combination of the rows of M' has at least k zero entries.
(4) There do not exist distinct subspaces Y and Z of $\mathcal{P}_q$ such that
   - *$dim(\langle Y, Z \rangle) = k$*
   - *$dim(Y) = dim(Z) = k-1$*
   - *$\langle Y, Z \rangle \subset \mathcal{O}_{k-1}$*
   - *$Y \cup Z \subset \mathcal{O}_{k-2}$*
   - *$Y \cap Z \subset \mathcal{O}_{k-3}$.*
(5) There is no integer s with $k < s \leq q$ such that the Reed–Solomon code $\mathcal{R}$ with entries in $\mathbb{F}_q$ of dimension s can have $s-k+2$ columns $\mathcal{B} = \{b_1, ..., b_{s-k+2}\}$ added to it, such that:
   - *Any $s \times s$ submatrix of $\mathcal{R} \cup \mathcal{B}$ containing the first $s-k$ columns of $\mathcal{B}$ is independent (non-zero determinant).*
   - *$\mathcal{B} \cup \{[0, 0, ..., 0, 1]^\top\}$ is independent.*

Then the following holds true:

(Part A) : (1), (2), (3) and (5) are equivalent.
(Part B) : (3) implies (4).

20

Covering of *SU*(2) and Quantum Lattices

A central question in quantum computing is how elements of $SU(2)$ (the collection of 2×2 unitary matrices with determinant 1) can be approximated by a small set of "generators". A classical bit is the basic unit of information used in classical computing, which has the states 1 or 0 [97]. Quantum computing extends this concept using the notion of quantum bits. Dirac notation is used to denote the basic states $|0\rangle$ and $|1\rangle$. Then a quantum bit, or qubit, is a pair of complex numbers α, β which correspond to the probability of the qubit being in the states $|0\rangle$ or $|1\rangle$. Thus, the quantum bit is represented as $\alpha|0\rangle + \beta|1\rangle$. Since α and β represent the probability of the qubit being in a particular state, it must hold that $|\alpha|^2+|\beta|^2 = 1$. Therefore, qubits can be represented by unit vectors in $\mathbb{C}^2$.

This idea extends to n-qubits [97]. An n-qubit is the tensor product

$$(\alpha_1|0\rangle + \beta_1|1\rangle) \otimes \cdots \otimes (\alpha_n|0\rangle + \beta_n|1\rangle)$$

As mentioned above, 1-qubits form the unit circle in $\mathbb{C}^2$, and it follows that n-qubits form vectors in $(\mathbb{C}^2)^{\otimes n}$. Examples of classical gates include the AND, OR, and NOT gates.

Since each 1-qubit is a unit vector, then 1-qubit quantum gates should take unit vectors to unit vectors. Thus, quantum gates are taken to have determinant 1. n-qubit gates are formed by tensoring 1-qubit matrices with the controlled NOT gate:

$$CNOT = \begin{bmatrix} 1 & 0 & 0 & 0 \\ 0 & 1 & 0 & 0 \\ 0 & 0 & 0 & 1 \\ 0 & 0 & 1 & 0. \end{bmatrix}$$

Near Extensions and Alignment of Data in $\mathbb{R}^n$: Whitney extensions of near isometries, shortest paths, equidistribution, clustering and non-rigid alignment of data in Euclidean space,
First Edition. Steven B. Damelin.

For the construction of quantum computers, a finite base set of quantum gates must be chosen so that they generate $SU(2)$. However, it turns out that it is not practical to consider this problem. The gate sets are constructed so that the elements they generate can approximate any quantum gate. Defining how a gate approximates another gate is half the battle, which usually entails constructing a metric-induced topology on $SU(2)$, but in general gate sets which generate dense subsets of $SU(2)$ are chosen. In a given context, a set which generates a dense subset of $SU(2)$ is referred to as a universal set in $SU(2)$, and by definition gives that it can approximate any gate in $SU(2)$ with arbitrary precision according to the chosen measure of approximation. However, as in all computing, the question becomes how to choose efficient universal sets to approximate elements in $SU(2)$. Efficient meaning that it requires the least amount of matrices (or some generalize notion of cost) to approximate all elements of $SU(2)$.

In this chapter, the goal is to construct a universal gate set $T*$ in $SU(2)$ that efficiently approximates all of $PSU(2)$ (the equivalence classes of $SU(2)$ under multiplication by -1) using a natural and simple notion of distance. Covering exponents constructed in [104] are used to measure the efficiency of a gate set in approximating every element of $SU(2)$. In general, $T*$ is constructed to minimize the maximal cost of approximating any gate. Our T^* can efficiently approximate $PSU(2)$ but does not quite efficiently approximate $SU(2)$. Our work follows very closely to that of [104]. The letter G is used to represent either $SU(2)$ and $PSU(2)$ (The projective special linear group). It is apparent from the context, and reiterated when necessary, which choice is being used.

20.1 Structure of $SU(2)$

Any element $M \in SU(2)$ can be written in terms of $\alpha, \beta \in \mathbb{C}$ as

$$\begin{bmatrix} \alpha & \beta \\ -\bar{\beta} & \bar{\alpha} \end{bmatrix}$$

Thus, M can be associated with some vector in $\mathbb{R}^4$. Note that

$$\det M = \alpha\bar{\alpha} + \beta\bar{\beta} = |\alpha|^2 + |\beta|^2 = 1$$

This relation allows points in $SU(2)$ to be identified with points on S^3. It is a powerful tool in computing the efficiency of universal sets of $SU(2)$. However, unlike S^3, $SU(2)$ does not have a standard topology by convention. Before notions of universality and closeness can be used, $SU(2)$ must be set up as a metric space with an induced topology. Define the distance between two matrices M, N as

$$d_G(M,N) = \sqrt{1 - \frac{|Tr(M^{\dagger}N)|}{2}} \tag{20.1}$$

where $M^{\dagger}$ represents the conjugate transpose of M.

This metric, induces a natural topology on the metric space (G, d_G) using balls $B_G(.)$ as open sets and using the Haar measure μ, see [104].

20.2 Universal Sets

Let Γ be a finite subset of G. The set Γ is said to be universal in G, with respect to this topology, if the subgroup of G generated by Γ is dense. If Γ is not universal, then there will be open balls that contain no elements generated by Γ.

We have [97, 104] for our purposes:

Theorem 20.12. *(Solovay–Kitaev) Let Γ be a finite universal set in $SU(2)$ and $\varepsilon > 0$. Then there exists a constant c such that for any $M \in SU(2)$, there is a finite product S of gates in Γ of length $O(\log^c(\frac{1}{\varepsilon}))$ such that $d_G(S, M) < \varepsilon$.*

Universality of Γ gives that any one matrix can be approximated with arbitrary precision. Theorem 20.12 gives that Γ can approximate $SU(2)$ with arbitrary efficiency and provides a estimation for the maximum length required to achieve this approximation. This theorem provides justification for studying the efficiency of universal gate sets in approximating all of $SU(2)$, instead of specific matrices. As computers are not typically constructed to perform single calculations, this is much more useful.

To consider the efficiency of a universal set, [104] provides an interesting idea. Fix a weight w on Γ (defined in [104]). Then $\forall \gamma \in \langle \Gamma \rangle$ defines the height of γ in Γ as

$$h(\gamma) = \min\left\{\sum_i w(c_i) \, : \, c_i \in \Gamma, \gamma = \prod c_i\right\} \tag{20.2}$$

Note that this notion of height is heavily dependent on the choice of w. Given such a weight w, define the following set for $t > 0$

$$V_\Gamma(t) = \{\gamma \in \langle \Gamma \rangle \, : \, h(\gamma) \leqslant t.\}$$

Let $\varepsilon > 0$. Define the covering length of Γ within ε, denoted t_ε as in [104], as follows

$$t_\varepsilon = \min\{t \in \mathbb{N} \, : \, G \subset \bigcup_{\gamma \in V_\Gamma(t)} B_G(\varepsilon)\} \tag{20.3}$$

The calculation of t_ε is the ultimate prize. Especially, if it can be computed or even bounded as a function of ε, then t_ε can provide an explicit measure of the cost of approximating $SU(2)$.

20.3 Covering Exponent

Let Γ be a universal set in G, and $\varepsilon > 0$. A calculation with the Haar measure μ (see [104]) shows that for any $t > 0$

$$|V_\Gamma(t_\varepsilon)|\,\mu(B_G(\varepsilon)) \geqslant 1. \tag{20.4}$$

If Γ approximates G optimally, then inequality (20.4) becomes an equality. In general, as $|V_\Gamma(t_\varepsilon)|$ becomes close to $\frac{1}{\mu(B_G(\varepsilon))}$, the overlap between the balls centered at points in $V_\Gamma(t_\varepsilon)$ is minimized. Thus, Γ becomes more efficient at approximating G. For a universal set Γ in G and a Haar measure μ on G, the covering exponent as given in [104] is defined as

$$K(\Gamma) = \limsup_{\varepsilon\to 0} \frac{\log|V_\Gamma(t_\varepsilon)|}{\log\left(\frac{1}{\mu(B_G(\varepsilon))}\right)}. \tag{20.5}$$

Note that K is heavily dependent on t_ε, and does vary with a choice of G. The second part is to be expected, $PSU(2)$ is almost half the elements of $SU(2)$ and should typically be easier to generate. The dependence of K on t_ε is more convenient.

20.4 An Efficient Universal Set in *PSU*(2)

What makes a universal set optimal, or even efficient in approximating $SU(2)$? In [104], there are several different ideas offered for what makes optimal choices of quantum gates to approximate $SU(2)$ (along with some properties useful to computer scientists). For this construction, let $G = PSU(2)$. The condition from [104] used to construct the efficient set T is that there exists a normal form over T, for which any matrix generated by T has a unique representation in this normal form. This allows $V_T(t)$ to be studied concretely for any $t > 0$ and the following theorem to be proved [104]

Theorem 20.13.

$$4/3 \leq K(T) \leq 2.$$

In our paper [74], an efficient universal set $T*$ over $PSU(2)$ is constructed via Pauli matrices, using a metric of the covering exponent. Then, the relationship between $SU(2)$ and S^3 is manipulated to correlate angles between points on S^3 and to give a sufficient condition on the maximum of angles $< . >$ between points on certain lattices. It is shown how this sufficient condition can be used to bound $K(T*)$ from above and below.

We have for $PSU(2)$ and dot product < ... > [74], the following theorem.

Theorem 20.14. *Let $a \in S^3$ be given. Suppose there exists $0 < \delta < 2/3$ with the following property: There is $b \in \mathbb{Z}^4$ and $k \in \mathbb{Z}$ with $|b| = 5^k$ and $\langle a, \frac{b}{5^k} \rangle \geqslant 1 - 5^{\frac{-k}{2-\delta}}$. Then*

$$4/3 \leq K(T*) \leq 2 - \delta.$$

A similar theorem holds using primes $p \equiv 1\ (mod\ 4)$.

21

The Unlabeled Correspondence Configuration Problem and Optimal Transport

The work below is motivated by the difficulty in trying to match point sets in the absence of labels in the sense that often one does not know which point to map to which. This is referred to commonly as the unlabeled problem. We discuss briefly optimal transport.

21.1 Unlabeled Correspondence Configuration Problem

In the paper [26], we investigate ways to align two point configurations by first finding a correspondence between points and then constructing a function which aligns the configurations. The terms reordering and relabeling are used interchangeably. Examples are given in [26] to show, for example, in $\mathbb{R}^2$, when in certain configurations, some distributions of distances do not allow good alignment and how we can partition certain configurations into polygons in order to construct maximum possible correspondences between these configurations, considering their areas. Algorithms are described for certain configurations with matching points along with examples where we find a permutation which gives us a relabeling, and also the required affine transformation which aligns certain configurations. See also Figures 21.1–21.2.

For what we present, we assume $n = 2$ and $m \geq 2$.

21.1.1 Non-reconstructible Configurations

For the definition below, we will call two finite subsets $P, Q \subset \mathbb{R}^2$ congruent if there exists an isometry $f : \mathbb{R}^2 \to \mathbb{R}^2$ with $f(P) = Q$. We then have:

Near Extensions and Alignment of Data in $\mathbb{R}^n$: Whitney extensions of near isometries, shortest paths, equidistribution, clustering and non-rigid alignment of data in Euclidean space,
First Edition. Steven B. Damelin.

Definition 21.1. *By a relabeling, we mean that if there is an initial labeling of ordered points in two congruent configurations, we reorder them in such a way that there is a correspondence between the points.*

An example: suppose we are given two configurations of five points and there is an initial labeling of ordered points $\{a, b, c, d, e\}$ in the first set and $\{a*, b*, c*, d*, e*\}$ in the second set where a corresponds to $a*$, b to $b*$, c to $e*$, d to $d*$ and e to $c*$, one such relabeling will come from the permutation $\binom{1\ 2\ 3\ 4\ 5}{1\ 2\ 5\ 4\ 3}$.

We have:

Proposition 21.1. *[17] Suppose $m \neq 4$. A permutation $f \in S_{\binom{m}{2}}$ is a relabeling if and only if for all pairwise distinct indices $i, j, k \in \{1, ..., m\}$ we have:*

$$f \cdot \{i, j\} \cap f \cdot \{i, k\} \neq \emptyset.$$

In other words, we need to take the edges of equal length between the two configurations we are considering and check if there is a mutual vertex between all such pairs for a given permutation $f \in S_{\binom{m}{2}}$. This permutation is what will give us the labeling if it does exist.

Question "QU"

In the context of our work, we consider two given m-point configurations $P := \{p_1, ..., p_m\}$ and $Q := \{q_1, ..., q_m\}$ with their corresponding pairwise distances $D_P = \{dp_{ij} | dp_{ij} = |p_i - p_j|, 1 \leq i, j \leq m\}$ and $D_Q = \{dq_{ij} | dq_{ij} = |q_i - q_j|, 1 \leq i, j \leq m\}$ with $D_P = D_Q$ up to some reordering and $|D_P| = |D_Q| = \binom{m}{2}$.

We then want to find if $\exists \{i, k\}, \{j, l\}$ such that $dp_{ik} = dq_{jl}$ $\forall i, j, k, l \in$ not true, we hope to disregard a certain number of *bad points* from both configurations in order to achieve this.

Let us call this this question "QU".

21.1.2 Example

Below is an example of question QU with two different 4-point configurations in $\mathbb{R}^2$ which have the same *distribution of distances*. The corresponding equal distances between the two configurations are represented in the same color, and we have two edges with distances $1, 2$ and $\sqrt{5}$, but it is obvious that there doesn't exist a Euclidean transformation corresponding to the two configurations.

From this example we can construct infinitely many sets of two different configurations with the same distribution of distances. This can be done by simply

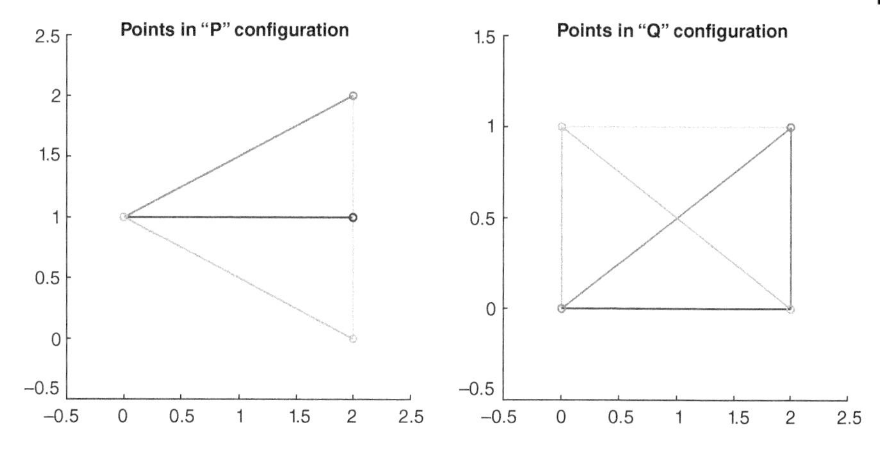

Figure 21.1 Two different 4-point configurations with the same distribution of distances.

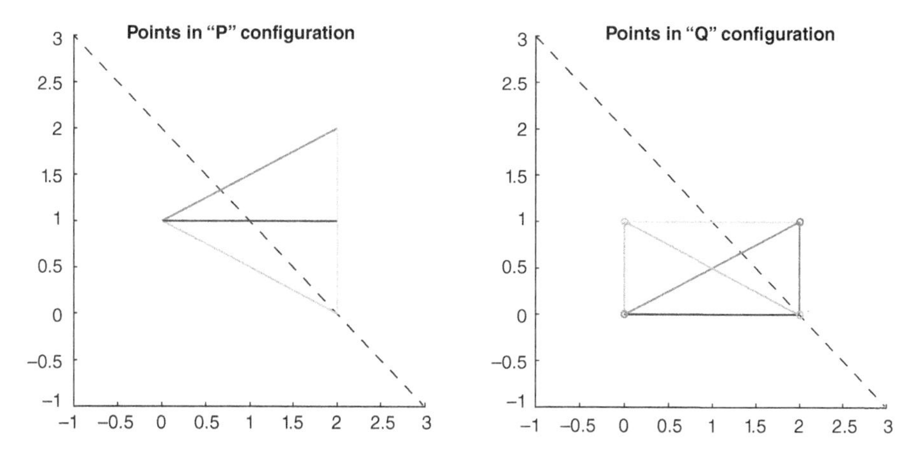

Figure 21.2 Configurations with the same distribution of distances for $m \geq 4$.

adding as many points as desired on the same location across the dashed line in both the configurations of Figures 21.1 and 21.2.

In this example, it suffices to exclude one point from the two configurations to be able to obtain a Euclidean motion to correspond them.

We note that if it suffices to exclude a single bad-point from both configurations P and Q, so that P and Q differ only by a Euclidean motion action, then iterating through the potential pairs of bad-points will take $\mathcal{O}(m^2)$ complexity. The issue still arises in determining whether the points we excluded results in two congruent configurations.

We explore question QU more.

21.1.3 Partition Into Polygons

We consider the following approach for question QU to see which points can be excluded if possible, from two configurations P and Q with some geometry. Our idea is to partition the configurations into polygons and compare polygons of the *same area*, in order to determine existing point correspondences between P and Q. For any subsets $\{i, j, ...\} \subseteq \{1, ..., m\}$ or $\{s, t\} \subseteq \{1, ..., k\}$ we consider in the upcoming sections, the elements of each to be distinct.

21.1.4 Considering Areas of Triangles—*10-step Algorithm*

Considering our two m-point configurations which we write as $P = \{p_1, ..., p_m\}$ and $Q = \{q_1, ..., q_m\}$, we partition them into a total of $\binom{m}{3}$ *triangles* and considering the distance between our three vertex points in each case, let us say indexed i, j, k, we have the distances $dp_{ij}, dp_{ik}, dp_{jk}$ and analogously $dq_{i'j'}, dq_{i'k'}, dq_{j'k'}$.

We now compute the areas as follows:

$$A_{ijk} = \sqrt{s(s - dp_{ij})(s - dp_{ik})(s - dp_{jk})}$$

$$\text{where } s := \frac{dp_{ij} + dp_{ik} + dp_{jk}}{2}$$

$$B_{i'j'k'} = \sqrt{s'(s' - dq_{i'j'})(s' - dq_{i'k'})(s' - dq_{j'k'})}$$

$$\text{where } s' := \frac{dq_{i'j'} + dq_{i'k'} + dq_{j'k'}}{2}$$

and consider the sets of areas

$$\mathcal{A} = \{A_{ijk} | \forall \{i, j, k\} \subseteq \{1, ..., m\}\}$$

and

$$\mathcal{B} = \{B_{i'j'k'} | \forall \{i', j', k'\} \subseteq \{1, ..., m\}\}$$

where $|\mathcal{A}| = |\mathcal{B}| = \binom{m}{3} = \frac{m(m-1)(m-2)}{6}$.

We further partition the above sets as follows:

$$\mathcal{A}_1 = \{A_{ijk} | A_{ijk} \in \mathcal{A} \text{ and } \exists B_{i'j'k'} \in \mathcal{B} \text{ s.t. } A_{ijk} = B_{i'j'k'}\}.$$

$$\mathcal{B}_1 = \{B_{i'j'k'} | B_{i'j'k'} \in \mathcal{B} \text{ and } \exists A_{ijk} \in \mathcal{A} \text{ s.t. } B_{i'j'k'} = A_{ijk}\}.$$

$$\mathcal{A}_2 = \mathcal{A} \backslash \mathcal{A}_1 \qquad \mathcal{B}_2 = \mathcal{B} \backslash \mathcal{B}_1.$$

Note that it may not be true that $|\mathcal{A}_1| = |\mathcal{B}_1|$, as the areas need not all be distinct. We essentially want the *shapes* formed by our points in the two sets which are *identical*. Let us assume our sets $\mathcal{A}_1$ and $\mathcal{B}_1$ are in ascending order with respect to the modes of the areas. We undertake the following steps in order to check which points to disregard and which permutations are valid:

- Disregard all points from from P and Q which are vertices of triangles in $\mathcal{A}_2$ and $\mathcal{B}_2$ respectively, but at the same time not vertices of any triangle in $\mathcal{A}_1$ and $\mathcal{B}_1$.
- In order we take $A_{ijk} \in \mathcal{A}_1$ and the corresponding triangles in $\mathcal{B}_1$, with $A_{ijk} = B_{i'j'k'}$.
- If the distances of the sides of the triangles corresponding to A_{ijk} and $B_{i'j'k'}$ don't match, disregard the triangles with area $B_{i'j'k'}$.
- If the distances match up, we assign the points the corresponding points from P to Q and essentially start constructing our permutation. So thus far we have

  $$\begin{pmatrix} i & j & k & \cdots \\ i' & j' & k' & \cdots \end{pmatrix}$$

  Alternatively, we can match the points between A_{ijk} and $B_{i'j'k'}$ which have the same corresponding *angles*.
- Note that we might have more than one possible permutation, so for now we keep track of all of them and list them as $\alpha_s^{(t)} = \begin{pmatrix} i & j & k & \cdots \\ i' & j' & k' & \cdots \end{pmatrix}$ for s being the indicator of the triangle we take from $\mathcal{A}_1$, and t being the indicator of the corresponding triangle in $\mathcal{B}_1$ in order (so if three triangles correspond, we have $t \in \{1, 2, 3\}$).
  - For triangles with three distinct inner angles we will have one permutation, for *isosceles* triangles two permutations, and for *equilateral* triangles $3! = 6$ permutations.
  - In the case of *squares* when considering quadrilaterals, we will have $4! = 24$ permutations which will be discussed more carefully in [26].
  - This can be thought of as *matching angles* between equidistant edges of our polygons.
- Go to the next triangle in $\mathcal{A}_1$ (which might have the same area as our previous triangle), and repeat steps 2–4.
  - If the distances of our current triangle match with those of our previous triangle, simply take all previous permutations and *concatenate* them. (We use

the the term *combine*.) So for example $\left(\alpha_1^{(1)}\right)_2 = \left(\alpha_1^{(1)}\alpha_1^{(2)}\right)$, where the index v in $\left(\alpha_s^{(t)}\right)_v$ indicates the combination we have with $\alpha_s^{(t)}$ being the first *element* of the permutation as above. We therefore get a total of $\prod_{\iota=0}^{\nu-1}(t-\iota)$ permutations we are currently keeping track of, where ν is the number of *elements* in the constructions thus far. Note that in the above procedure we assume no common points, and the case where mutual points exist is described below.

- If our current and previous triangles *share points*, we consider the combination of two triangles in $\mathcal{B}_1$ with the same corresponding areas and shapes and matching points as the combination of the two triangles taken from $\mathcal{A}_1$, so we will either get a quadrilateral (if they share two points) or two triangles sharing an vertex (not a pentagon), and check whether all $\binom{4}{2}$ or $\binom{5}{2}$ distances between our two shapes match up. If they do, we replace the permutations we are keeping track of, and disregard any permutations from before which do not satisfy the conditions of this bullet point.
- If our current and previous triangles *don't share points*, we essentially repeat steps 2–5 and *repeat* the permutations we are keeping track of, in a similar manner to that shown in step 6.
- At this point we have traversed through all triangles in both $\mathcal{A}_1$ and $\mathcal{B}_1$ with the same area, and have constructed permutations (not necessarily all of the same size) which can be considered as *sub-correspondence* of points between P and Q (meaning that more points may be included to the correspondences). We are now going to be considering the triangles with area of the next lowest mode and repeat steps 2–8, while keeping track of the permutations we have thus far. Some steps though will be slightly modified as now we are considering various "shapes" (corresponding to our permutations), and in the above steps when referring to our *previous triangle*, we will now be considering our *previous shapes*.
- Repeating the above until we traverse through all triangles in $\mathcal{A}_1$ and $\mathcal{B}_1$ will give us a certain number of permutations, and for our problem we can simply take the permutations of the largest size (we might have multiple) and the points which are not included in that permutation can be considered as *bad points* for the problem. Note that certain points might be considered as *bad* for certain permutations and not for others, which depends entirely on the configurations P and Q.

A Brief Explanation of the Above Approach

The idea of the above approach is to disregard non-identical shapes and configurations of the point sets P and Q, while simultaneously constructing the desired permutations of *sub-configurations* which have the same shape. Note that we begin with the triangle areas which have the smallest mode in order to simplify the implementation of this algorithm.

We guess that at least one framework to study smooth extensions in a setting of small distorted pairwise distances will require that the maximum permutations constructed have size n, where all points will be included. A drawback of this approach, is that we keep track of a relatively large number of permutations through this process, but when going through each set of triangles of the same area, a lot of them are disregarded.

21.1.5 Graph Point of View

Another way to view this problem is as a graph problem, where our points correspond to vertices and the distances correspond to weighted edges between the vertices of a fully-connected graph. Considering the two graphs G_P and G_Q constructed by P and Q respectively, our goal is to find existing *subgraph isomorphisms*. This is known to be an *NP-complete* problem. The triangles we were using above will correspond to 3-node cliques, while quadrilaterals will correspond to 4-node cliques.

Considering this idea will actually make the problem (problem (1)) significantly easier to implement, by taking advantage of the adjacency of matrices of the two graphs. In [26] we use this concept to find the correspondence between the points, when it is known that there exists one for all points in P and Q. This is the *graph isomorphism* problem, and belongs in the *NP-intermediate* complexity class. We do not present it here.

21.1.6 Considering Areas of Quadrilaterals

Specifically aiming now at Problem (1) we find the following idea interesting.

We can partition $P = \{p_1, \ldots, p_m\}$ and $Q = \{q_1, \ldots, q_m\}$, by partitioning them into a total of $\binom{m}{4}$ *quadrilaterals*, and consider the $\binom{4}{2} = 6$ distances between our four vertex points in each case. If we take four distinct points indexed i, j, k, l, we have the set of distances $\mathcal{DP}_{ijkl} = \{dp_{ij}, dp_{ik}, dp_{il}, dp_{jk}, dp_{jl}, dp_{kl}\}$ and analogously $\mathcal{DQ}_{i'j'k'l'} = \{dq_{i'j'}, dq_{i'k'}, dq_{i'l'}, dq_{j'k'}, dq_{j'l'}, dq_{k'l'}\}$.

We now compute the areas as follows: define the following:

$$r := \max\{d \in \mathcal{DP}_{ijkl}\}, s := \max\{d \in \mathcal{DP}_{ijkl} \backslash \{r\}\}$$

$$\{a, b, c, d\} := \mathcal{DP}_{ijkl} \backslash \{r, s\},$$

where a, c correspond to distances of edges which do not share a vertex

$$A_{ijkl} = \frac{1}{4}\sqrt{4r^2s^2 - (a^2 + c^2 - b^2 - d^2)^2}$$

$$r' := \max\{d \in \mathcal{DQ}_{i'j'k'l'}\},\ s' := \max\{d \in \mathcal{DQ}_{ijkl} \backslash \{r'\}\}$$

$$\{a', b', c', d'\} := \mathcal{DQ}_{i'j'k'l'} \backslash \{r', s'\},$$

where a';c' correspond to distances of edges which do not share a vertex

$$B_{i'j'k'l'} = \frac{1}{4}\sqrt{4r'^2 s'^2 - (a'^2 + c'^2 - b'^2 - d'^2)^2}$$

Consider the sets of areas

$$\mathcal{A} = \{A_{ijkl} | \forall \{i, j, k, l\} \subseteq \{1, ..., m\}\}.$$

$$\mathcal{B} = \{B_{i'j'k'l'} | \forall \{i', j', k', l'\} \subseteq \{1, ..., m\}\}.$$

Here, $|\mathcal{A}| = |\mathcal{B}| = \binom{m}{4} = \frac{m(m-1)(m-2)(m-3)}{24}$.

We further partition the above sets as follows:

$$\mathcal{A}_1 = \{A_{ijkl} | A_{ijkl} \in \mathcal{A} \text{ and } \exists B_{i'j'k'l'} \in \mathcal{B} \text{ s.t. } A_{ijkl} = B_{i'j'k'l'}\}.$$

$$\mathcal{B}_1 = \{B_{i'j'k'l'} | B_{i'j'k'l'} \in \mathcal{B} \text{ and } \exists A_{ijkl} \in \mathcal{A} \text{ s.t. } B_{i'j'k'l'} = A_{ijkl}\}.$$

$$\mathcal{A}_2 = \mathcal{A} \backslash \mathcal{A}_1 \qquad \mathcal{B}_2 = \mathcal{B} \backslash \mathcal{B}_1.$$

We can now follow the same algorithm described in Section 21.1.4, with the exception that now we will be considering four points at a time, rather than three. Depending on the point-configurations P and Q, either this approach or the previous approach might be more efficient, but this cannot be determined a priori.

21.1.7 Partition Into Polygons for Small Distorted Pairwise Distances

The purpose of this section is to see now what can be done related to possible frameworks for Problem (1) using this circle of ideas.

21.1.8 Areas of Triangles for Small Distorted Pairwise Distances

Our sets will have the following property we will call "EP":

$$(1 - \varepsilon_{ij}) \leq \frac{dp_{ij}}{dq_{i'j'}} \leq (1 + \varepsilon_{ij}),\ \forall \{i, j, i', j'\} \subseteq \{1, ..., m\},\ i \neq j,\ i' \neq j'$$

Here, $\varepsilon_{ij} > 0$ are small enough.

Property EP replaces: $dp_{ij} = dq_{i'j'}, \forall \{i, j, i', j'\} \subseteq \{1, ..., m\}$.

Let us state, Theorem 21.1.1 below as an interesting result (although quite computational).

Theorem 21.1.1. *For our usual setup, it holds that for three points in our two point configurations P and Q with indices and areas* $\{i, j, k\}$, A_{ijk} *and* $\{i', j', k'\}$, $B_{i'j'k'}$ *respectively, the following holds: given* $\varepsilon_{ij} > 0$ *small enough. Then property EP holds iff*

$$\sqrt{(B_{i'j'k'})^2 - \frac{1}{4} \cdot H_1} \leq A_{ijk} \leq \sqrt{(B_{i'j'k'})^2 + \frac{1}{4} \cdot H_2}$$

where H_1, H_2 *depend on* $E := max\{\varepsilon_{st} | \{s, t\} \subseteq \{i, j, k\}\}$, *and the elements of the distribution of distances of* $B_{i'j'k'}$.

Proof. Considering the area of the triangles defined by the points p_i, p_j, p_k and the corresponding points q_i', q_j', q_k', we define $\varepsilon_{ij-} := (1 - \varepsilon_{ij})$, $\varepsilon_{ij+} := (1 + \varepsilon_{ij})$, and obtain the following three inequalities for each triangle:

$$dq_{i'j'} \cdot \varepsilon_{ij-} \leq dp_{ij} \leq dq_{i'j'} \cdot \varepsilon_{ij+}.$$

$$dq_{i'k'} \cdot \varepsilon_{ik-} \leq dp_{ik} \leq dq_{i'k'} \cdot \varepsilon_{ik+}.$$

$$dq_{j'k'} \cdot \varepsilon_{jk-} \leq dp_{jk} \leq dq_{j'k'} \cdot \varepsilon_{jk+}.$$

In order to simplify our computations, we define:

$$E := max\{\varepsilon_{st} | \{s, t\} \subseteq \{i, j, k\}\}$$

and

$$E_- := (1 - E) \qquad E_+ := (1 + E)$$

We then have:

$$dq_{s't'} \cdot E_- \leq dp_{st} \leq dq_{s't'} \cdot E_+$$

for all pairs

$$\{s, t\} \subseteq \{i, j, k\}$$

and the following:

$$s := dp_{ij} + dp_{ik} + dp_{jk}$$

$$s' := dq_{i'j'} + dq_{i'k'} + dq_{j'k'}$$

It then follows that for all pairs $\{s, t\} \subseteq \{i, j, k\}$, that

$$(2dq_{s't'}) \cdot E_- \leq 2dp_{st} \leq (2dq_{s't'}) \cdot E_+$$

and

$$(-2dq_{s't'}) \cdot E_+ \leq -2dp_{st} \leq (-2dq_{s't'}) \cdot E_-$$

which give

$$(dq_{i'j'}+dq_{i'k'}+dq_{j'k'})\cdot E_- \leq dp_{ij}+dp_{ik}+dp_{jk} \leq (dq_{i'j'}+dq_{i'k'}+dq_{j'k'})\cdot E_+$$

and so

$$s' \cdot E_- \leq s \leq s' \cdot E_+$$

which means that

$$(s' \cdot E_- - 2dq_{i'j'} \cdot E_+) \leq (s - 2dq_{ij}) \leq (s' \cdot E_+ - 2dq_{i'j'} \cdot E_-)$$

Taking advantage of the *triangle inequality*, $s' \leq 2dq_{i'j'}$, we get the following *bounds*

$$\begin{aligned} 2dq_{i'j'} \cdot (E_- - E_+) &\leq (s' \cdot E_- - 2dq_{i'j'} \cdot E_+) \leq (s - 2dq_{ij}) \\ &\leq (s' \cdot E_+ - 2dq_{i'j'} \cdot E_-) \leq 2s' \cdot (E_+ - E_-) \\ 2dq_{i'j'} \cdot (E_- - E_+) &\leq (s - 2dq_{ij}) \leq 2s' \cdot (E_+ - E_-) \\ (-4E) \cdot dq_{i'j'} &\leq (s - 2dq_{ij}) \leq (4E) \cdot s' \\ 0 &\leq (s - 2dq_{ij}) \leq (4E) \cdot s' \end{aligned}$$

We know that the area of the triangle defined by the points in the configurations P and Q are respectively

$$\begin{aligned} A_{ijk} &= \sqrt{\frac{s\left(\frac{s}{2} - dp_{ij}\right)\left(\frac{s}{2} - dp_{ik}\right)\left(\frac{s}{2} - dp_{jk}\right)}{2}} \\ &= \frac{1}{2}\sqrt{s(s - 2dp_{ij})(s - 2dp_{ik})(s - 2dp_{jk})} \end{aligned}$$

and so

$$A_{ijk} = \frac{1}{2} \cdot \sqrt{S} \text{ for } S := s(s - 2dp_{ij})(s - 2dp_{ik})(s - 2dp_{jk})$$

$$\begin{aligned} B_{i'j'k'} &= \sqrt{\frac{s'\left(\frac{s'}{2} - dq_{i'j'}\right)\left(\frac{s'}{2} - dq_{i'k'}\right)\left(\frac{s'}{2} - dq_{j'k'}\right)}{2}} \\ &= \frac{1}{2}\sqrt{s'(s' - 2dq_{i'j'})(s' - 2dq_{i'k'})(s' - 2dq_{j'k'})} \end{aligned}$$

and so

$$B_{i'j'k'} = \frac{1}{2} \cdot \sqrt{S'} \text{ for } S' := s'(s' - 2dq_{i'j'})(s' - 2dq_{i'k'})(s' - 2dq_{j'k'})$$

We do not undertake any simplifications, and from the above inequalities considering the indices $\{i, j, k\}$ and $\{i', j', k'\}$, we perform the following computations with suitable symbols for ease of analysis.

$$s' \prod_{\iota' \neq \kappa'} (s' \cdot E_- - 2dq_{\iota'\kappa'} \cdot E_+) \leq s \prod_{\iota \neq \kappa} (s - 2dp_{\iota\kappa})$$

$$\leq s' \prod_{\iota' \neq \kappa'} (s' \cdot E_+ - 2dq_{\iota'\kappa'} \cdot E_-)$$

$$s' \prod_{\iota' \neq \kappa'} [(s' - 2dq_{\iota'\kappa'}) - (s' + 2dq_{\iota'\kappa'}) \cdot E] \leq s \prod_{\iota \neq \kappa} (s - 2dp_{\iota\kappa})$$

$$\leq s' \prod_{\iota' \neq \kappa'} [(s' - 2dq_{\iota'\kappa'}) + (s' + 2dq_{\iota'\kappa'}) \cdot E]$$

$$s'(\alpha_1 - \beta_1)(\alpha_2 - \beta_2)(\alpha_3 - \beta_3) \leq s \prod_{\iota \neq \kappa} (s - 2dp_{\iota\kappa})$$

$$\leq s'(\alpha_1 + \beta_1)(\alpha_2 + \beta_2)(\alpha_3 + \beta_3)$$

$$s'\Big[\alpha_1\alpha_2\alpha_3 - [\alpha_3\beta_2(\alpha_1 - \beta_1) + \alpha_1\beta_3(\alpha_2 - \beta_2) + \alpha_2\beta_1(\alpha_3 - \beta_3)] - \beta_1\beta_2\beta_3\Big]$$

$$\leq s \prod_{\iota \neq \kappa} (s - 2dp_{\iota\kappa}) \leq$$

$$\leq s'\Big[\alpha_1\alpha_2\alpha_3 + [\alpha_3\beta_2(\alpha_1 + \beta_1) + \alpha_1\beta_3(\alpha_2 + \beta_2) + \alpha_2\beta_1(\alpha_3 + \beta_3)] + \beta_1\beta_2\beta_3\Big]$$

$$S' - s'\Big[\alpha_3\beta_2(\alpha_1 - \beta_1) + \alpha_1\beta_3(\alpha_2 - \beta_2) + \alpha_2\beta_1(\alpha_3 - \beta_3) + \beta_1\beta_2\beta_3\Big] \leq S \leq$$

$$\leq S' + s'\Big[\alpha_3\beta_2(\alpha_1 + \beta_1) + \alpha_1\beta_3(\alpha_2 + \beta_2) + \alpha_2\beta_1(\alpha_3 + \beta_3) + \beta_1\beta_2\beta_3\Big]$$

$$S' - H_1 \leq S \leq S' + H_2$$

Comparing the areas of two corresponding triangles from the two point-configurations we then get:

$$4 \cdot (B_{i'j'k'})^2 - H_1 \leq 4 \cdot (A_{ijk})^2 \leq 4 \cdot (B_{i'j'k'})^2 + H_2$$

$$(B_{i'j'k'})^2 - \frac{1}{4} \cdot H_1 \leq (A_{ijk})^2 \leq (B_{i'j'k'})^2 + \frac{1}{4} \cdot H_2$$

$$\sqrt{(B_{i'j'k'})^2 - \frac{1}{4} \cdot H_1} \leq A_{ijk} \leq \sqrt{(B_{i'j'k'})^2 + \frac{1}{4} \cdot H_2}$$

This gives the required upper and lower bounds. ■

21.1.9 Considering Areas of Triangles (part 2)

For areas of triangles with property EP, we construct the sets of areas of the partitioned triangles as follows:

$$\mathcal{A} = \{A_{ijk} | \forall \{i, j, k\} \subseteq \{1, ..., m\}\}$$

$$\mathcal{B} = \{B_{i'j'k'} | \forall \{i', j', k'\} \subseteq \{1, ..., m\}\}$$

$$\mathcal{A}_1 = \{A_{ijk} | A_{ijk} \in \mathcal{A} \text{ w/ } E \text{ and } \exists B_{i'j'k'} \in \mathcal{B}, \text{s.t. } |\sqrt{(B_{i'j'k'})^2 - \frac{H_1}{4}}| \leq A_{ijk} \leq |\sqrt{(B_{i'j'k'})^2 + \frac{H_2}{4}}|\}$$

$$\mathcal{B}_1 = \{B_{i'j'k'} | B_{i'j'k'} \in \mathcal{B} \text{ and } \exists A_{ijk} \in \mathcal{A} \text{ w/ } E, \text{s.t. } |\sqrt{(B_{i'j'k'})^2 - \frac{H_1}{4}}| \leq A_{ijk} \leq |\sqrt{(B_{i'j'k'})^2 + \frac{H_2}{4}}|\}$$

$$\mathcal{A}_2 = \mathcal{A} \backslash \mathcal{A}_1 \qquad \mathcal{B}_2 = \mathcal{B} \backslash \mathcal{B}_1$$

Here, we use suitable constants when needed.

We then follow exactly the same *10-step algorithm* as in Section 21.1.4 to get the desired result although now it is very unlikely that two or more triangles will have exactly the same area.

21.1.10 Areas of Quadrilaterals for Small Distorted Pairwise Distances

Let us state, Theorem 21.1.2 below as an interesting result.

Theorem 21.1.2. *For our usual setup, it holds that for four points in our two point configurations P and Q with indices and areas $\{i, j, k\}$, A_{ijk} and $\{i', j', k'\}$, $B_{i'j'k'}$ respectively, the following holds: Given $\varepsilon_{ij} > 0$ small enough, then property EP holds iff*

$$\sqrt{(B_{i'j'k'l'})^2 \cdot (1 + E^2)^2 - \frac{\hat{H}_2}{16}} \leq A_{ijkl} \leq \sqrt{(B_{i'j'k'l'})^2 \cdot (1 + E^2)^2 + \frac{\hat{H}_2}{16}}$$

where $\hat{H}_1, \hat{H}_2$ depend on $E := max\{\varepsilon_{st} | \{s, t\} \subseteq \{i, j, k, l\}\}$, and the elements of the distribution of distances of $B_{i'j'k'l'}$.

Proof. As before, we perform the following computations with suitable symbols for ease of analysis.

We consider our two m-point configurations $P = \{p_1, ..., p_m\}$ and $Q = \{q_1, ..., q_m\}$, and partition them into a total of $\binom{m}{4}$ *quadrilaterals* , and take into account $\binom{4}{2} = 6$ distances between our four points in each case. If we take the four points indexed i, j, k, l, we have the set of distances $\mathcal{DP}_{ijkl} = \{dp_{ij}, dp_{ik}, dp_{il}, dp_{jk}, dp_{jl}, dp_{kl}\}$ and analogously $\mathcal{DQ}_{i'j'k'l'} = \{dq_{i'j'}, dq_{i'k'}, dq_{i'l'}, dq_{j'k'}, dq_{j'l'}, dq_{k'l'}\}$ for our second configuration. We also define $\varepsilon_{ij-} := (1-\varepsilon_{ij})$, $\varepsilon_{ij+} := (1+\varepsilon_{ij})$, and get the following six inequalities for each triangle:

$$dq_{i'j'} \cdot \varepsilon_{ij-} \leq dp_{ij} \leq dq_{i'j'} \cdot \varepsilon_{ij+} \qquad dq_{j'k'} \cdot \varepsilon_{jk-} \leq dp_{jk} \leq dq_{j'k'} \cdot \varepsilon_{jk+}$$
$$dq_{i'k'} \cdot \varepsilon_{ik-} \leq dp_{ik} \leq dq_{i'k'} \cdot \varepsilon_{ik+} \qquad dq_{j'l'} \cdot \varepsilon_{jl-} \leq dp_{jl} \leq dq_{j'l'} \cdot \varepsilon_{jl+}$$
$$dq_{i'l'} \cdot \varepsilon_{il-} \leq dp_{il} \leq dq_{i'l'} \cdot \varepsilon_{il+} \qquad dq_{k'l'} \cdot \varepsilon_{kl-} \leq dp_{kl} \leq dq_{k'l'} \cdot \varepsilon_{kl+}$$

Following a similar approach to what was shown previously, we define the following parameters and compute the areas:

$$E := max\{\varepsilon_{st} | \{s,t\} \subseteq \{i,j,k,l\}\}$$
$$E_- := (1-E) \qquad E_+ := (1+E)$$

which gives

$$dq_{s't'} \cdot E_- \leq dp_{st} \leq dq_{s't'} \cdot E_+ \text{, for all pairs } \{\text{s,t}\} \subseteq \{i,j,k,l\}$$

and

$$r := \max\{d \in \mathcal{DP}_{ijkl}\} \qquad s := \max\{d \in \mathcal{DP}_{ijkl} \backslash \{r\}\}$$
$$\{a,b,c,d\} := \mathcal{DP}_{ijkl} \backslash \{r,s\},$$

where a, c correspond to distances of edges which do not share a vertex

$$S := (a^2 + c^2 - b^2 - d^2) \qquad \tilde{S} := (a^2 + b^2 + c^2 + d^2)$$
$$A_{ijkl} = \frac{1}{4}\sqrt{4r^2s^2 - (a^2 + c^2 - b^2 - d^2)^2}$$

and so

$$A_{ijkl} = \frac{1}{4}\sqrt{4r^2s^2 - S^2}$$
$$r' := \max\{d \in \mathcal{DQ}_{i'j'k'l'}\} \qquad s' := \max\{d \in \mathcal{DQ}_{i'j'k'l'} \backslash \{r'\}\}$$
$$\{a',b',c',d'\} := \mathcal{DQ}_{i'j'k'l'} \backslash \{r',s'\},$$

where $a';c'$ correspond to distances of edges which do not share a vertex

$$S' := (a'^2 + c'^2 - b'^2 - d'^2) \qquad \tilde{S}' := (a'^2 + b'^2 + c'^2 + d'^2)$$
$$B_{i'j'k'l'} = \frac{1}{4}\sqrt{4r'^2s'^2 - (a'^2 + c'^2 - b'^2 - d'^2)^2} \Longrightarrow B_{i'j'k'l'} = \frac{1}{4}\sqrt{4r'^2s'^2 - S'^2}$$

It then follows that for all pairs $\{s,t\} \subseteq \{i,j,k\}$

$$(dq_{s't'})^2 \cdot (E_-)^2 \leq (dp_{st})^2 \leq (dq_{s't'})^2 \cdot (E_+)^2$$
$$-(dq_{s't'})^2 \cdot (E_+)^2 \leq -(dp_{st})^2 \leq -(dq_{s't'})^2 \cdot (E_-)^2$$

which implies that

$$(r's')^2 \cdot (E_-)^4 \leq (rs)^2 \leq (r's')^2 \cdot (E_+)^4$$

and

$$\begin{aligned}
&\left[(a'^2 + c'^2) \cdot (E_-)^2 - (+b'^2 + d'^2) \cdot (E_+)^2\right] \leq (a^2 + c^2 - b^2 - d^2) \\
&\quad \leq \left[(a'^2 + c'^2) \cdot (E_+)^2 - (+b'^2 + d'^2) \cdot (E_-)^2\right] \\
&\left[(a'^2 + c'^2) \cdot (1 - 2E + E^2) - (+b'^2 + d'^2) \cdot (1 + 2E + E^2)\right] \\
&\quad \leq (a^2 + c^2 - b^2 - d^2) \leq \\
&\leq \left[(a'^2 + c'^2) \cdot (1 + 2E + E^2) - (+b'^2 + d'^2) \cdot (1 - 2E + E^2)\right] \\
&\left[(a'^2 + c'^2 - b'^2 - d'^2) \cdot (1 + E^2) - 2E \cdot (a'^2 + c'^2 + b'^2 + d'^2)\right] \\
&\quad \leq (a^2 + c^2 - b^2 - d^2) \leq \\
&\leq \left[(a'^2 + c'^2 - b'^2 - d'^2) \cdot (1 + E^2) + 2E \cdot (a'^2 + c'^2 + b'^2 + d'^2)\right] \\
&\left[S' \cdot (1 + E^2) - \tilde{S}' \cdot (2E)\right] \leq S \leq \left[S' \cdot (1 + E^2) + \tilde{S}' \cdot (2E)\right] \\
&-\left[S' \cdot (1 + E^2) + \tilde{S}' \cdot (2E)\right]^2 \leq -S^2 \leq -\left[S' \cdot (1 + E^2) - \tilde{S}' \cdot (2E)\right]^2 \\
&-S'^2 \cdot (1 + E^2)^2 - \left[\tilde{S}' \cdot (2E) \cdot [2E\tilde{S}' + S'(1 + E^2)]\right] \leq -S^2 \leq \\
&\leq -S'^2 \cdot (1 + E^2)^2 - \left[\tilde{S}' \cdot (2E) \cdot [2E\tilde{S}' - S'(1 + E^2)]\right] \\
&-S'^2 \cdot (1 + E^2)^2 - H_1 \leq -S^2 \leq -S'^2 \cdot (1 + E^2)^2 - H_2
\end{aligned}$$

Comparing the areas of two corresponding quadrilaterals from the two point-configurations we then get:

$$\begin{aligned}
&4(r's')^2 \cdot (E_-)^4 - S'^2 \cdot (1 + E^2)^2 - H_1 \leq 4(rs)^2 - S^2 \\
&\quad \leq 4(r's')^2 \cdot (E_+)^4 - S'^2 \cdot (1 + E^2)^2 - H_2 \\
&4(r's')^2 \cdot [(1 + E^2)^2 - 4E(1 - E + E^2)] - S'^2 \cdot (1 + E^2)^2 - H_1 \\
&\quad \leq 4(rs)^2 - S^2 \leq \\
&\leq 4(r's')^2 \cdot [(1 + E^2)^2 + 4E(1 + E + E^2)] - S'^2 \cdot (1 + E^2)^2 - H_2 \\
&\left[4(r's')^2 - S'^2\right] \cdot (1 + E^2)^2 - \left[16(r's')^2 \cdot (E - E^2 + E^3) + H_1\right] \\
&\quad \leq 4(rs)^2 - S^2 \leq
\end{aligned}$$

$$\leq \left[4(r's')^2 - S'^2\right] \cdot (1 + E^2)^2 + \left[16(r's')^2 \cdot (E + E^2 + E^3) - H_2\right]$$

$$\left[4(r's')^2 - S'^2\right] \cdot (1 + E^2)^2 - \hat{H}_1 \leq 4(rs)^2 - S^2$$

$$\leq \left[4(r's')^2 - S'^2\right] \cdot (1 + E^2)^2 + \hat{H}_2$$

$$(B_{i'j'k'l'})^2 \cdot (1 + E^2)^2 - \frac{\hat{H}_1}{16} \leq (A_{ijkl})^2 \leq B^2_{i'j'k'l'} \cdot (1 + E^2)^2 + \frac{\hat{H}_2}{16}$$

$$\sqrt{(B_{i'j'k'l'})^2 \cdot (1 + E^2)^2 - \frac{\hat{H}_1}{16}} \leq A_{ijkl} \leq \sqrt{(B_{i'j'k'l'})^2 \cdot (1 + E^2)^2 + \frac{\hat{H}_2}{16}}$$

Finally, we have for this section the following. ■

21.1.11 Considering Areas of Quadrilaterals (part 2)

For areas of quadrilaterals with property EP we construct the sets of areas of the partitioned quadrilaterals as follows.

$$\mathcal{A} = \left\{A_{ijkl} | \forall \{i, j, k, l\} \subseteq \{1, ..., m\}\right\}$$

$$\mathcal{B} = \left\{B_{i'j'k'l'} | \forall \{i', j', k', l'\} \subseteq \{1, ..., m\}\right\}$$

$$\mathcal{A}_1 = \left\{A_{ijkl} | A_{ijkl} \in \mathcal{A} \text{ w/ } E \text{ and } \exists B_{i'j'k'l'} \in \mathcal{B}, \text{ s.t. } | \right.$$

$$\sqrt{(B_{i'j'k'l'})^2 \cdot (1 + E^2)^2 - \frac{\hat{H}_1}{16}} | \leq$$

$$\left. \leq A_{ijkl} \leq | \sqrt{(B_{i'j'k'l'})^2 \cdot (1 + E^2)^2 + \frac{\hat{H}_2}{16}} | \right\}$$

$$\mathcal{B}_1 = \left\{B_{i'j'k'l'} | B_{i'j'k'l'} \in \mathcal{B} \text{ and } \exists A_{ijkl} \in \mathcal{A} \text{ w/ } E, \text{ s.t. } | \right.$$

$$\sqrt{(B_{i'j'k'l'})^2 \cdot (1 + E^2)^2 - \frac{\hat{H}_1}{16}} | \leq$$

$$\left. \leq A_{ijkl} \leq | \sqrt{(B_{i'j'k'l'})^2 \cdot (1 + E^2)^2 + \frac{\hat{H}_2}{16}} | \right\}$$

$$\mathcal{A}_2 = \mathcal{A} \backslash \mathcal{A}_1 \qquad \mathcal{B}_2 = \mathcal{B} \backslash \mathcal{B}_1$$

We then follow a similar *10-step algorithm* to get the desired result although now it is very unlikely that two or more quadrilaterals will have exactly the same area. We use suitable constants when needed as before.

In[26], we study further topics, for example of reconstruction from distances, relabeling, visualization, and algorithms in great detail using much machinery for example the Kabsch algorithm.

22

A Short Section on Optimal Transport

The Euclidean distance between points in point clouds might not be a reliable measure for proximity if, for instance, the point clouds are not approximately aligned or the point clouds are noisy. To establish better correspondences, many have looked into improving the similarity measures by defining features/descriptors for point cloud data that not only capture the location of a point, but also encode the local geometry of the object in the vicinity of that point (e.g., curvature). These methods then rely on a nearest neighbor matching in the feature space instead of the raw input space. Moreover, to obtain robustness and avoid false correspondences, these methods often use the Random Sample Consensus (RANSAC) algorithm or its variations. Unfortunately, however, RANSAC significantly increases the computational cost of the registration algorithm as it requires running the correspondence problem multiple times for different random subsets of the data [6].

While extremely successful for some applications, most nonlinear dimension reduction methods embedd data in Euclidean space assuming that Euclidean distances between data points are geometrically and pratically meaningful. Also, the manifold hypothesis is often assumed. See Chapter 4. In vision classification and signal identification translated copies of a single image for example can have large Euclidean distance even though they are semantically identical and should be assigned the same label. To this end, no-collision distances have been carried out with Wasserstein (OT) and linearized Wasserstein distances (LOT), respectively by various authors as well as Procrustes analysis with Wasserstein distances.

Solving the correspondence between points in the source and target point clouds is closely related to the celebrated optimal transport (OT) problem. In short, for two empirical distributions and given a transportation cost, the OT problem seeks the optimal assignment between the samples of the two distributions such that the expected transportation cost between the assigned samples is minimized. Hence, one can treat the point clouds as empirical distributions and solve an OT

Near Extensions and Alignment of Data in $\mathbb{R}^n$: Whitney extensions of near isometries, shortest paths, equidistribution, clustering and non-rigid alignment of data in Euclidean space,
First Edition. Steven B. Damelin.

problem to find the correspondences between the points in the source and target point clouds. This principle has led to many OT-based point-cloud registration algorithms. However, a major limitation of OT is the mass preservation assumption, which translates to requiring all points in the source to be matched to all the points in the target point cloud. The mass preservation assumption limits the application of OT to problems where the points must be partially matched, e.g., noisy or occluded point clouds and, in general, partial registration problems. Various ideas have been recently developed to allow the application of OT for partial registration. For instance, some dynamically infer the mass of each particle (i.e., importance of a point in the point cloud) so that the OT problem can ignore the particles with zero/small mass. Others have looked into defining outlier bins and solving the OT problem while allowing for matching points to the outlier bin. This latter idea is deeply rooted in the Optimal Partial Transport (OPT) problem, which allows for creation and destruction of mass [6].

In our paper [6], we use OPT and its sliced variation as an excellent match for partial and robust registration problems and demonstrate its performance in various registration problems.

The main idea in our paper [6] is studying robust parametric non-rigid registration frameworks using OPT frameworks in $\mathbb{R}^d$ using a sliced OPT framework for accelerated non-rigid registration between large scale point clouds. It would be interesting to study these frameworks in a setting with Slides and Slow Twists.

23

Conclusion

This monograph provides fascinating connections between several problems which lie on the interface of algebraic geometry, approximation theory, computer vision, data science, differential geometry, harmonic analysis, applied harmonic analysis, manifold and machine learning, networks, optimal transport, partial differential equations, probability, shortest paths, quantum analysis, neuroscience and signal processing.

The work leads to many questions which I hope will inspire future research.

Near Extensions and Alignment of Data in $\mathbb{R}^n$: Whitney extensions of near isometries, shortest paths, equidistribution, clustering and non-rigid alignment of data in Euclidean space,
First Edition. Steven B. Damelin.

References

[1] T. Amir, S. Kovalsky and N. Dym, *Symmetrized robust procrustes: Constant-factor approximation and exact recovery*, arxiv 2207.08592.

[2] M. Andrade, M. P Gomez-Carracedo, W. Krzanowski and M. Kubista, *Procrustes rotation in analytical chemistry, a tutorial*, Chemometrics and Intelligent Laboratory Systems, **72** (2) (2004), pp. 123–132.

[3] J. H. Ann, S. B. Damelin, P. Bigeleisen, *Medical image segmentation using modified Mumford segmentation methods*, Ultrasound-Guided Regional Anesthesia and Pain Medicine, eds P. Bigeleisen, Chapter 40, Birkhauser, 2009.

[4] M. Artetxe, G. Labaka and E. Agirre, *A robust self-learning method for fully unsupervised cross-lingual mappings of word embeddings*, Proceedings of the 56th Annual Meeting of the Association for Computational Linguistics (Volume 1: Long Papers), pp. 789–798, Melbourne, Australia, July 2018. Association for Computational Linguistics.

[5] Y. Bai, B. Schmitzer, M. Thorpe and S. Kolouri, *Sliced optimal transport*, arXiv:2212.08049.

[6] Y. Bai, T. Huy, S. B. Damelin and S. Kolouri, *Noisy point cloud registration via optimal partial transport*, preprint.

[7] B. Bajnok, S. B. Damelin, J. Li and G. Mullen, *A constructive method of scattering points on d-dimensional spheres using finite fields*, Computing, **68** (2002), pp. 97–109.

[8] E. Begelfor and M. Werman, *Affine invariance revisited*, 2006 IEEE Computer Society Conference on Computer Vision and Pattern recognition.

[9] E. Bengelfor and M. Werman, *The world is not always flat or learning curved manifolds*, preprint.

[10] C. Bocci, E. Carlini and J. Kileel, *Hadamard products of linear spaces*, Journal of Algebra **448** (2016), pp. 595–617.

Near Extensions and Alignment of Data in $\mathbb{R}^n$: Whitney extensions of near isometries, shortest paths, equidistribution, clustering and non-rigid alignment of data in Euclidean space,
First Edition. Steven B. Damelin.

[11] S. Bora, S. B. Damelin, D. Kaiser and J. Sun, *An algebraic-coding equivalence to the maximal coding separable conjecture*, arxiv 1705.06136, Involve to appear.

[12] S. Ball and M. Lavrauw, *Arcs in finite projective spaces*, arXiv:1908.10772v1.

[13] S. Ball, *The Grassl-Rotteler cyclic and consta-cyclic MDS codes are generalized Reed-Solomon codes*, arXiv:2112.11896v2.

[14] S. Ball, G. Gamboa and M. Lavrauw, *On additive MDS codes over small fields*, arXiv:2012.06183v1.

[15] M. Boutin and P. Bazin, *Structure from motion: A new look from the point of view of invariant theory*, SIAM Journal on Applied Mathematics, **64** (4) (2004), pp. 1156–1174.

[16] M. Boutin and G. Kemper, *Which point sets are determined by the distribution of their pairwise distances*? International Journal of Computational Geometry and Applications **17** (1) (2007), pp. 31–43.

[17] M. Boutin and G. Kemper, *On reconstructing n-point configurations from the distributions of distances or areas*, arxiv 0304192.

[18] P. Boyd and L. Vandenberghe, *Convex Optimization*, Cambridge university press, 2004.

[19] M. Boyer, Y. Lipman, E. St Clair, J. Puente, B. A. Patel, T. Funkhouser, J. Jernvall and I. Daubechies, *Algorithms to automatically quantify the geometric similarity of anatomical surfaces*, Proceedings of the National Academy of Sciences, **108** (45) (2011), pp. 18221–18226.

[20] P. Callahan, *Dealing with Higher Dimensions: The Well-Separated Pair Decomposition and Its Applications*, PhD thesis, John Hopkins University, 1995, pp. 1–179.

[21] E. J. Candès, *The restricted isometry property and its implications for compressed sensing*, C. R. Math, **346** (9) (2008), pp. 589–592.

[22] E. J. Candès, *Compressive sampling*, Proceedings of the International of Congress Mathematicians, Madrid, Spain, August 2006.

[23] E. J. Candes, J. K. Romberg and T. Tao, *Stable signal recovery from incomplete and inaccurate measurements*, Communications on Pure and Applied Mathematics: A Journal Issued by the Courant Institute of Mathematical Sciences, **59** (8) (2006), pp. 1207–1223.

[24] K. A. Cawse, S. B. Damelin, R. McIntyre, M. Mitchley, L. du Plessis and M. Sears, *An investigation of data compression for hyperspectral core image data*, Proceedings of the Mathematics in Industry Study Group 2008, South Africa, (2008), 1–25.

[25] K. Cawse, S. B. Damelin, A. Robin and M. Sears, *A parameter free approach for determining the intrinsic dimension of a hyperspectral image using random matrix theory*, IEEE Transaction on Image Processing, **22** (4) (2013), pp. 1301–1310.

[26] N. Charalambides, S. B. Damelin and B. Swartz, *On isometries and equivalences between sets of points: labeled and unlabeled points*, arXiv:1705.06146.

[27] S. B. Damelin, *On Whitney extensions, Whitney extensions of small distortions and Laguerre polynomials*, arXiv:2302.08045.

[28] S. B. Damelin, *A walk-through energy, discrepancy, numerical integration and group invariant measures on measurable subsets of Euclidean space*, Numerical Algorithms, **48** (1–3) (2008), pp. 213–235.

[29] S. B. Damelin, *Minimal discrete energy problems and numerical integration on compact sets in Euclidean spaces*, Algorithms for Approximation, (2007), pp. 369–378.

[30] S. B. Damelin, *The Hilbert transform and orthonormal expansions for exponential weights*, In: Chui, S. and Stoekler (eds.) Approximation Theory X: Abstract and Classical Analysis, Vanderbilt Univ. Press (2002), pp. 117–135.

[31] S. B. Damelin, *The distribution of general interpolation arrays for exponential weights*, Electronic Transactions of Numerical Analysis, **12** (2002), pp. 12–20.

[32] S. B. Damelin, *On bounds for diffusion functions, discrepancy and fill distance metrics*, Springer Lecture Notes in Computational Science and Engineering, **58** (2008), pp. 32–42.

[33] S. B. Damelin, *Advances on regularity and dislocation properties of energy minimizing sets, discrepancy, manifold learning and their applications*, Algorithms for Approximation, (2007), pp. 369–400.

[34] S .B. Damelin and K. Diethelm, *Weighted polynomial approximation and Hilbert transforms: their connections to the numerical solution of singular integral equations*, Proceedings of Dynamic Systems and Applications, **4** (2004), Ed. G. S. Ladde, N.G. Medhin. M. Sambandham, pp. 20–26.

[35] S. B. Damelin and K. Diethelm, *Interpolatory product quadratures for Cauchy principal value integrals with Freud weights*, Numerical Mathematics, **83** (1999), pp. 87–105.

[36] S. B. Damelin and K. Diethelm, *Boundedness and uniform approximation of the weighted Hilbert transform on the real line*, Numerical Functional Analysis and Optimization, **22** (1 and 2) (2001), pp. 13–54.

[37] S. B. Damelin and K. Diethelm, *Numerical solution of Fredholm integral equations on the line*, Journal of Integral equations and Applications, **13** (3) (2004), pp. 273–292.

[38] S. B. Damelin and K. Diethelm, *Analytic and numerical analysis of singular Cauchy integrals with exponential-type weights*, arXiv 1711.09495.

[39] S. B. Damelin and C. Fefferman, *On smooth Whitney extensions of almost isometries with small enough distortion, interpolation and alignment in $\mathbb{R}^D$-part 1*, arXiv:1411.2451.

[40] S. B. Damelin and C. Fefferman, *On the Whitney extension-interpolation-alignment problem for almost isometries with small enough distortion in* $\mathbb{R}^D$, arXiv:1411.2468.

[41] S. B. Damelin and C. Fefferman, *On the Whitney distortion extension problem for* $C^m(\mathbb{R}^n)$ *and* $C^\infty(\mathbb{R}^n)$ *and its applications to interpolation and alignment of points in* $\mathbb{R}^n$, arXiv:1505.06950.

[42] S. B. Damelin, C. Fefferman and W. Glover, *A BMO theorem for* ε*-distorted diffeomorphisms from* $\mathbb{R}^D$ *to* $\mathbb{R}^D$ *with applications to manifolds of speech and sound*, Involve, **5-2** (2012), pp. 159–172.

[43] S. B. Damelin and P. Grabner, *Energy functionals, Numerical integration and asymptotic equidistribution on the sphere*, Journal of Complexity, 19 (2003), pp. 231–246. (Postscript) Corrigendum, Journal of Complexity, (20) (2004), pp. 883–884.

[44] S. B. Damelin, Y. Gu, D. Wunsch and R. Xu, *Fuzzy adaptive resonance theory, diffusion functions and their applications to clustering and bi-clustering*, Mathematical Modelling of Natural Phenomena, **3** (10) (2015), pp. 206–211.

[45] S. B. Damelin and K. Hamm, Reordered weighted orthornormal expansions on the real line, (in preparation).

[46] S. B. Damelin, F. Hickernell, D. Ragozin and X. Zeng, *On energy, discrepancy and G invariant measures on measurable subsets of Euclidean space*, Journal of Fourier Analysis and its Applications **6** (2010), pp. 813–839.

[47] S. B. Damelin and N. Hoang, *On surface completion and image inpainting by biharmonic functions: Numerical aspects*, International Journal of Mathematics and Mathematical Sciences, **5** (2018), pp. 1–8. Implementation in the scikit-image package, scikit-image.

[48] J. Sun, S. B. Damelin, D. Kaiser and S. Bora, *An algebraic-coding equivalence to the maximal coding separable conjecture*, Involve to appear, arXiv:1705.06136.

[49] S. B. Damelin, J. Levesley, D. L. Ragozin and X. Sun, *Energies, Group Invariant Kernels and Numerical Integration on Compact Manifolds*, Journal of Complexity, **25** (2009), pp. 152–162.

[50] S. B. Damelin and V. Maymeskul, *Minimal Discrete Energy Problems and Numerical Integration on Compact Sets in Euclidean Spaces*, Algorithms for Approximation, (2007) pp. 359–368.

[51] S. B. Damelin and V. Maymeskul, *On Point Energies, Separation Radius and Mesh Norm for s-Extremal Configurations on Compact Sets in* $\mathbb{R}^n$, Journal of Complexity, **21** (6) (2006), pp. 845–863.

[52] S. B. Damelin and D. S. Lubinsky, Necessary and sufficient conditions for mean convergence of Lagrange interpolation for Erdős weights, Candian Mathematics of Journal, **40** (1996), pp. 710–736.

[53] S. B. Damelin and W. Miller, *Mathematics of signal processing*, Cambridge Texts in Applied Mathematics, **48**, February 2012.

[54] S. B. Damelin, G. Mullen and G. Michalski, *The cardinality of sets of k-independent vectors over finite fields*, Monatshefte fur Mathematik, **150** (2008), pp. 289–295.

[55] S. B. Damelin, G. Mullen, G. Michalski and D. Stone, *On the number of linearly independent binary vectors of fixed length with applications to the existence of completely orthogonal structures, Monatshefte fur Mathematik*, **1** (2003), pp. 1–12.

[56] S. B. Damelin, D. L. Ragozin and M. Werman, *On min-max affine approximants of convex or concave real-valued functions from $\mathbb{R}^k$, Chebyshev equioscillation and graphics*. In: Hirn, M., Li, S., Okoudjou, K.A., Saliani, S. (eds.) Excursions in Harmonic Analysis. Applied and Numerical Harmonic Analysis, vol. 6. Springer, Cham (2021). doi.org/10.1007/978-3-030-69637-5-19.

[57] D. L. Donoho, *Compressed sensing*, IEEE Transactions on Information Theory, **52** (4) (2006), pp. 1289–1306.

[58] J. Draisma, E. Horobet, G. Ottaviani, B. Sturmfels and R. Thomas, *The Euclidean distance degree of an algebraic variety*, Foundations of Computational Mathematics, 16: 99–149, 2016.

[59] N. Dym, *Exact Recovery with Symmetries for the Doubly-Stochastic Relaxation*, SIAM Journal on Applied Algebra and Geometry (SIAGA), 2018.

[60] N. Dym, https://nadavdym.github.io/.

[61] N. Dym and Y. Lipman, *Exact Recovery with Symmetries for Procrustes Matching*, SIAM Journal on Optimization, (2017).

[62] N. Dym and S. Z. Kovalsky, *Linearly converging quasi branch and bound algorithms for global rigid registration*, Proceedings of the IEEE/CVF International Conference on Computer Vision, (2019), pp. 1628–1636.

[63] C. Fefferman, *Whitney's extension problems and interpolation of data*, Bulletin of the American Mathematical Society (N.S), **46** (2) (2009), 207–220.

[64] C. Fefferman and A. Israel, *Fitting Smooth Functions to Data*, CBMS Regional Conference Series in Mathematics, 135, 2019.

[65] G. Floystad, J. Kileel and G. Ottaviani, *The Chow form of the essential variety in computer vision*, Journal of Symbolic Computation, **86**.

[66] D. Hahnel, S. Thrun and W. Burgard, *An extension of the ICP algorithm for modeling nonrigid objects with mobile robots*, IJCAI-03, Proceedings of the Eighteenth International Joint Conference on Artificial Intelligence, (2018), pp. 97–119.

[67] K. Hamm, https://github.com/stevendamelinkeatonhamm/damelinhamm slowtwistsslides.

[68] T. Hastie, R. Tibshirani and J. Friedman, *The Elements of Statistical Learning*, Springer (2001).

[69] Q.H. Huang, B. Adams, M. Wicke and L. J. Guibas, *Non-rigid regsitration under isometric deformations*, Eurographics Symposium on Geometry Processing, (2008).

[70] S. J. Hwang, S. B. Damelin and A. O. Hero III, *Shortest path through random points*, The Annals of Applied Probability, **26** (5) (2016), 2791–2823.

[71] F. John, *Rotation and strain*, Communications of Pure and Applied Mathematics, **14** (3) (1961), pp. 391–413.

[72] F. John and L. Nirenberg, *On functions of bounded mean oscillation*, Communications on Pure and Applied Mathematics, **14** (1961), pp. 415–426.

[73] G. Kalsi and S. B Damelin, *Preprocessing power weighted shortest path data using a s-Well Separated Pair Decomposition*, arXiv:2103.11216.

[74] A. Green and S. B. Damelin, *On the approximation of the quantum gates using lattices*, arXiv:1506.05785.

[75] J. Kileel, *Algebraic geometry for computer vision*, PhD thesis, University of Berkeley, 2017.

[76] J. Kileel, *Subspace power method for symmetric tensor decomposition and generalized PCA*, arXiv:1912.04007.

[77] J. Kileel, A. Bandeira, B. Blum-Smith, A. Perry, J. Weed and A. Wein, *Estimation under group actions: recovering orbits from invariants*, arXiv:1712.10163.

[78] J. Kileel and K. Kohn, *Snapshot of algebraic vision*, arXiv.2210.11443.

[79] J. Kileel, A. Moscovich, N. Zelesko and A. Singer, *Earthmover-based manifold learning for analyzing molecular conformation spaces*, arXiv:1911.06107.

[80] J. Kileel, Z. Kukelova, T. Pajdla and B. Sturmfels, *Distortion varieties*, Foundations of Computational Mathematics **18** (2018), pp. 1043–1071.

[81] J. Kileel, A. Moscovich, N. Zelesko and A. Singer, *A Manifold learning with arbitrary norms*, arXiv: 2012.14172.

[82] N. Kimia, A. Durmus, L. Chizat, S. Kolouri, S. Shahrampour and U. Simsekli, *Statistical and topological properties of sliced probability divergences*, Advances in Neural Information Processing Systems **33** (2020): 20802–20812.

[83] A. Kolpakov and M. Werman, *Robust affine feature matching via quadratic assignment on Grassmannians*, arXiv:2303.02698.

[84] A. Kolpakov and M. Werman, *An approach to robust ICP initialization*, arXiv:2212.05332.

[85] S. Kolouri, S. R. Park, M. Thorpe, D. Slepcev and G.K. Rohde. *Optimal mass transport: Signal processing and machine-learning applications*, IEEE Signal Processing Magazine, **34** (4) (2017), pp. 43–59.

[86] S. Kolouri, K. Nadjahi, U. Simsekli, R. Badeau, R and G. Rohde, *Generalized sliced Wasserstein distances*, Advances in Neural Information Processing Systems, **32** (2019).

[87] S. Kolouri, P. Pope, C. E. Martin and G. K. Rohde, *Sliced Wasserstein auto-encoders*, In International Conference on Learning Representations, (2019).

[88] S. Kolouri, K. Nadjahi, K., S. Shahrampour and Şimşekli, U, *Generalized Sliced Probability Metrics*, In: ICASSP 2022-2022 IEEE International Conference on Acoustics, Speech and Signal Processing (ICASSP), pp. 4513–4517. IEEE. [best paper].

[89] L. Kuipers and H. Niederreiter, *Uniform distribution of sequences*, Dover Publications, 1974, ISBN:0486450198.

[90] R. Lidl and H. Niederreiter, *Finite fields and their applications*, Encyclopedia of Mathematics and its Applications, Series Number 20, 2nd Edition.

[91] Y. Lipman and T. Funkhouser, *Mobius voting for surface correspondence*, In Proceedings SIGGRAPH, **72** (2009), pp. 1–72:12.

[92] N. Lubin, J. Goldberger and Y. Goldberg. *Aligning vector-spaces with noisy supervised lexicon*, Proceedings of the 2019 Conference of the North American Chapterof the Association for Computational Linguistics: Human Language Technologies, Volume 1(Long and Short Papers), pp. 460–465, Minneapolis, Minnesota, June 2019. Association for Computational Linguistics.

[93] H. Maron, N. Dym, I. Kezurer, S. Kovalsky and Y. Lipman, *Point Registration via Efficient Convex Relaxation*, ACM SIGGRAPH, (2016).

[94] L. G. Maxim, J. I. Rodriguez and B. Wang, SIAM Journal on Applied Algebra and Geometry **4** (1) (2020), pp. 28–48.

[95] D. McKenzie and S. B. Damelin, *Power weighted shortest paths for clustering Euclidean points*, Foundations of data science (American Institute of Mathematical Sciences), **1** (3) (2019), pp. 35–50.

[96] F. P. Medina, L. Ness, M. Weber and K. Y. Djima, *Heuristic framework for multiscale testing of the multi-manifold hypothesis*, Research in data science, Springer AWM Series, (2019), E. Gasparovic and Carotta Domeniconi, editors, ISBN-10: 3030115658.

[97] M. Nielson and I. Chuang, *Quantum Computation and Quantum Information*, Cambridge University Press, 2000.

[98] P. J. Olver, *Invariant signatures for recognition and symmetry*, I.M.A., University of Minnesota, April, 2006.

[99] I. Omer and M. Werman, *The bottleneck geodesic: Computing pixel affinity*, 2006 IEEE Computer Society Conference on Computer Vision and Pattern Recognition.

[100] M. Ovsjanikov, https://www.lix.polytechnique.fr/ maks/publications.html.

[101] M. Ovsjanikov, M. Ben-Chen, J. Solomon, A. Butscher and L. Guibas, *Functional functions: a flexible representation of functions between shapes*, ACM Transactions on Graphics (TOG), **31** (4) (2012), pp. 30:1–30:11.

[102] O. Peleg and M. Werman, *Fast and robust earth mover's distances*, 2009 IEEE 12th International Conference on Computer Vision, pp. 460–467.

[103] R. Raich, J. Costa, S.B. Damelin and Alfred O. Hero, *Classification constrained dimensionality reduction*, arXiv:0802.2906.

[104] P. Sarnak, *Letter on the Solovaey-Kitaev Theorem and Golden Gates.*

[105] P. H Schönemann, *A generalized solution of the orthogonal procrustes problem*, Psychometrika, **31** (1) (1966), pp. 1–10.

[106] J. Shanyu, J. Kolla and B. Shiffman, *A global Lojasiewicz inequality for algebraic varieties*, Transactions of the American Mathematical Society, **329** (2), pp. 813–818.

[107] Z. Shen, J. Feydy, P. Liu, A. H. Curiale, R. S. J. Estepar, R. S. J. Estepar and M. Niethammer, *Accurate point cloud registration with robust optimal transport*, NeurIPS, (2022).

[108] R. Singleton, *Maximum Distance Q-Nary Codes*, IEEE Transactions on Information Theory, **10** (1964), pp. 116–118.

[109] J. Sun, D. Capodilupo, S. B. Damelin, S. Freeman, M. Hua and M. Yu, *An Analytic and Probabilistic Approach to the Problem of Matroid Representibility*, arXiv:506.06425.

[110] J. B. Tenenbaum, V. Silva and J. C. Langford, *A global geometric framework for nonlinear dimensionality reduction*, Science, **290** (5500) (2000), pp. 2319–2323.

[111] S. Umeyama, Least squares estimation of transformations parameters between two point patterns, IEEE Transactions of Pattern Analysis and Machine Intelligence, **13** (4) (1991).

[112] J. H. Wells and L. R. Williams, *Embeddings and extensions in analysis*, Ergebnisse der Mathematik und iher Grenzgebietex, 84, Springer-Verlag, New York-Heidelberg, 1975.

[113] M. Werman and D. Weinshall, *Similarity and affine invariant distances between 2D point sets*, IEEE Transactions on Pattern Analysis and Machine Intelligence, **17** (8), 810–814.

[114] H. Whitney, *Analytic extensions of differentiable functions defined in closed sets*, Transactions of the American Mathematical Society **36** (1934), pp. 63–89.

[115] H. Whitney, *Differentiable functions defined in closed sets*, Transactions of the American Mathematical Society **36** (1934), pp 369–389.

[116] R. Xu, S. B. Damelin, D. C. Wunsch II, *Clustering of high-dimensional gene expression points with feature filtering methods and diffusion functions*, in Bio Medical Engineering and Informatics, BMEI, **1** (2008), pp. 245–249.

[117] R. Xu, S. B. Damelin and D. C. Wunsch II, *Applications of diffusion functions in gene expression data-based cancer diagnosis analysis*, In Proceedings of the 29th Annual International Conference of IEEE Engineering in Medicine and Biology Society, Lyon, France, August 2007, pp. 4613–4616.

[118] Z. M. Wang, N. Xue, L. Lei and G. Xia, *Partial Wasserstein adversarial network for non-rigid set registration*, ICLR, 2022.

[119] H. Yang, J. Shi and L. Carlone, *Teaser: Fast and certifiable point cloud registration*, IEEE Transactions on Robotics, **37** (2) (2020), 314–333.

[120] Q. Z, Zhou, J. Park and V. Koltun, *Fast global registration*, In: European Conference on Computer Vision, pp. 766–782. Springer, 2016.

Index

a

affine 3, 4, 6, 52, 57, 71, 73–74, 78, 92, 100, 102, 131
algebraic geometry 55–56, 89, 123, 149
algorithm 4, 6, 29, 30, 38, 119, 121, 131, 134, 137–138, 142, 145, 147, 148
alignment xiii, xiv, xix, 1, 2, 4–7, 109, 131
alternating tensor norm 54
Archimedean 35
artificial intelligence 29

b

bi-Lipchitz xiv, xvii, 114
ball 13, 43, 49, 61–63, 71, 79, 81, 89, 92–94, 96–97, 99, 115, 127–128
bounded xiii, xx, 1, 27, 37, 51, 61, 63, 94, 117
bounded mean oscillation, BMO xiii, xx, 61, 62
BMO norm 61

c

$C^m(\mathbb{R}^n)$ 1, 90
$C^\infty(\mathbb{R}^n)$ 1, 90
camera rotation 4, 23
cantor set 115
constant K viii, xviii, 25, 51–52, 55, 71, 74–75, 81–89
constant η for hyperplane viii, ix, 51–55, 59–60, 72–79, 84–87
η block ix, 71, 74, 76, 78, 82–83, 85–87
constant τ for separation viii, 51–52, 56, 59–60, 74–78, 83–84
center of mass 23
circle 19–20, 32, 109–110, 125
clustering viii, xiii, xx, 29, 33–34, 35, 37–39, 47, 82, 109
code x, xiv, 109, 121, 123–124
combinatorial design x, xiv, 109, 120, 121, 122

Near Extensions and Alignment of Data in $\mathbb{R}^n$: Whitney extensions of near isometries, shortest paths, equidistribution, clustering and non-rigid alignment of data in Euclidean space,
First Edition. Steven B. Damelin.

compact xiv, xvii, xviii, xix, xxi, 33–57, 89, 93, 109–110, 114–115, 121
complete graph 35, 37
compressed sensing viii, 29
computer vision xv, 4, 5, 7, 23, 56, 109, 149
configuration x, xi, xiv, xx, 31, 34, 36, 109–118, 131–144
constrained dimensionality reduction 30
continuous 1, 26, 34, 38, 57, 118
continuous density 36–37, 11
continuum limit-shortest paths viii, xiii, xx, 35–36
convex ix, 56–57
covering exponent 128
covers of $SU(2)$ 125
Cramer's rule 73
curse of dimensionality 29, 33

d

data xiii, xiv, xv, xix, xx, 1, 4, 6–7, 23, 29, 30, 33, 38–39, 109, 147, 149
deep learning 29, 30
deformation 4, 6
density 36–37, 111
distorted, distortion, near distortion xvii, 1, 2, 9, 25, 78, 89, 91, 95, 99, 103
distorted diffeomorphism vii, viii, ix, xiii, 9, 23–24, 43–63, 71–79
diffusion maps 29–30, 34
Dijkstra's algorithm 38
dimension reduction 30, 34, 117
discrepancy xiii, xiv, xx, 34–35, 109, 112, 117–118, 120–122
discrete/discrete set 7, 110–111
distance xiii, xiv, xvii, xx, 4, 6, 13, 18, 31–33, 35–38, 52, 54, 56, 95, 97, 109, 112, 114, 116, 123, 126, 134, 147
distortion 15
distribution xiii, xiv, xvii, xx, 4, 6, 13, 18, 31–38, 52, 54, 56, 95, 97, 109, 112, 114, 116, 123, 126, 134, 147

e

ellipsoid 32, 115
embedding 30, 31, 129
emperical distributions 147
energy xiii, xiv, xx, 34, 35, 109, 110–112, 118
equidistribution xiii, xiv, xx, 34–35, 109–112, 118
equivalence relation 39
Euclidean distance/norm xvii, xix, 2, 4, 36, 56, 147
Euclidean motion (improper/proper) xvii, xix, xx, 2, 3, 4, 9, 23–25, 41–44, 52–63, 71–86, 91, 93, 99, 103, 107, 133
extremal configurations 31, 34, 109–116

f

Fast Twist 13–15
feature space, feature descriptor 147
Fekete points 110–111
finite field xiii, xx, 109, 119–121
finite set xiii, xviii, xix, xx, 25, 27, 31, 33, 34, 36, 51–52, 59–60,

71, 74–78, 111, 114, 120, 123, 126–127, 131
finiteness principal 76
flat torus, torus 31, 110, 113–115, 118

g

global, globally rigid, global minimum, global Lojasiewtiz xviii, 23, 24, 31, 57, 112
gluing 78
graphics 4, 5, 7
group (video) 4
group invariance 34–35, 112, 117–118

h

Haar measure 127–128
Hausdorf measure 110, 112, 114, 117
hierarchical clustering 39
homogeneous 34, 118, 121
hyperplane 3, 27, 51–54, 57, 59, 71, 74, 89
hyperplane distance V_d p: 52, 54, 59, 74–75, 78
hyperspectral, hyperspectral image 23, 29, 34

i

image 23, 29, 34. 114, 147
induction 46, 82–83, 86
injectivity radius 31
Ind(k, n) 122
inner product space 123
interpolant, interpolation xiii, xix, 1, 2, 7, 57, 112
invariance 34–35, 112, 118
inverse function theorem 94
invertible 3, 74
invertible affine 3
Isomap 30
isometric, isometry, isometric embedding xiii, xviii, xix, xx, 2–3, 7, 12, 26, 29–32, 53
iterated closest point (ICP) 4
iterated slow twist 15

j

John Nirenburg inequality xx, xiii, 61–62
Johnson-Lindenstrauss xiv, xx, 29–30

k

Kabsch algorithm 145
kernel 34, 37–38, 118–119

l

$L_p(.)$, $1 \leq p \leq \infty$ 5, 30, 31
$l_p(.)$, $1 \leq p \leq \infty$ 5
label, labeled allgnment, labeling, labeling triangle x, xix, xx, 2, 5, 131, 132, 145, 147
Laplacian eigenfunctions 30
lattice 117
linear multidimensional scaling (MDS) 30
linear independence, linear independent xiii, xiv, xx, 109, 121, 122, 124
linear function/ transformation 3, 32, 73
linearized Wasserstein distances (LOT) 147
local, locally rigid xviii, 23–24, 31

local linear embedding 30
Lojasiewicz's inequality 55–57, 77
longest-leg path distance 37–38

m

manifold 30
manifold learning 30
maximal distance separable conjecture (MDS) xiii, xiv, xx, 109, 121, 123, 124
matroid 109, 121
mesh norm 110, 112, 114, 116
metric, metric space, metric tensor, norm xix, xvii, 2, 5, 30, 35–36, 56, 61–62, 125–128
minimal energy, energy, energy functional, Riesz energy xiii, xiv, xx, 34–35, 109–111

n

near alignment xix
near distortion xiii, xvii, xx, 1, 91, 95, 103, 105
near isometry xiii, xvii, xix, xx, 6, 7, 12, 26, 29, 31
near isometric embedding 29, 30, 31
near reflection, reflection xiii, 3, 51–53, 59, 87
nearest neighbor 4, 38
Newtonian, Newtonian kernel 34, 35, 36, 119
Newtonian configurations xiv, xx, 109–114
neural net 34
neuroscience xv, 4, 29, 109
non-linear 3, 6
non-linear deformation 6
non-linear trasformation 6
non-reconstructible configurations 131
non-rigid xiii, xvi, 1, 4–7, 148

o

$O(d)$, orthogonal group xiii, xx, 5, 26, 31, 61–63, 66–67
object and visual recognition 3–4
open, open set xiv, xix, 56, 89, 93–94, 118, 127
optimal partial transport (OPT), optimal transport (OT) xiv, xx, xv, 29, 109, 131, 147–149
optimization, convex/non-convex/min-max ix, 5–6, 23, 56–57
orthogonal projection 32, 54
overdetermined system 67

p

$PSU(d)$, projective unitary group 125–129
pairwise distance 2, 30, 43–44, 49, 77, 81, 132, 137–138, 142
permutation 3, 131–132, 135–138
parallelepiped 54, 115
parametric 6, 148
partition 33–34, 39, 45, 47, 81, 85, 95, 131, 134–135, 137–138, 141, 142, 145
partition of unity xiv, 95, 103
polygon 134, 138
polynomial 109, 120–123

positive definite, positive semi-definite, negative semi-definite 9, 34
power 29, 35–37, 112, 123, 126
probability density 35–37
probability/measure/convergence c.c 35–37
prime 119–124, 129
principal component analysis (PCA) 30
Procrustes 1, 3–5 23
Procrustes: Wasserstein 147
p-wspm allgorithm 38
Projective, Projective orthogonal transformation 4, 26, 30, 32, 54

q

quadrilateral, quadrilateral algorithm 135–137, 142, 145
quantum lattice xiii, xiv, xx, 125, 128
quasi smooth arcs 115

r

$\mathbb{R}^n$ xiii, xiv, xvii, xviii
$\mathbb{R}^d$ xiii, xiv, xvii, xviii
random 35–36, 21, 28, 30, 109, 117, 147
RANSAC algorithm 147
reach 31–32
rectifiable 34
Reed Solomon code 123, 124
registration 4, 6, 23, 147–148
regularized distance 95
Riemannian metric 35–36
restricted isometry 30–31
rigid xiii, xvii, 1, 4–6, 23–24
robotics 4, 5
rotation 4, 10–11, 12, 20, 54, 72, 110–111, 118
rotational norm on alternating tensors 54

s

$SO(d)$, special orthogonal group xvii, 4, 10, 20, 23
$SU(d)$, special unitary group xiv, 125–128
separation (packing) radius 110, 114, 116
shape, surface 3, 6–8, 56, 135–136
shortest paths, shortest path clustering xiii, xv, xx, 29–30, 34–38, 109, 148
Signal processing xv, 23, 29–30, 107
similarity, similarity kernel, similarity function, similarity measure 3, 34, 38, 147
simplex 25, 27, 51
singular value decomposition 30
sliced optimal partial transport (OPT) 148
Slide 11, 15–22, 52, 54
Slow Twist 10–13, 20–21, 149
smooth-smooth set, smooth denisity, smooth map xix, xx, 6, 9–11, 29, 37–38, 45, 49, 56, 59, 62, 89, 91, 94–95, 100, 103, 105, 115, 137
sphere 21–22, 25–26, 34, 45, 110, 112, 115, 117–118, 121

spherical t-design 120
symmetric 61

t

tensor, tensor product, wedge product xiii, 35, 51–54
translation xvii, 118
triangle 134, 138, 141

u

universal, universal set 126–129

v

variety 56
volume measure 52, 61–62

w

Wasserstein metric, Wasserstein (OT) 5, 146
Whitney cube xiv, x, 95–96, 99
Whitney extension, near Whitney extension, Whitney machinary xiii, xiv, xix, xvi, 1, 9, 29, 32, 77, 95